EMPOWERED TO DECEIVE

Book 2

J.B. Kingsley-Lauren

EMPOWERED TO DECEIVE
First published in Australia by Jean Thompson 2018

National Library of Australia Cataloguing-in-Publication entry:

Creator: Kingsley-Lauren, J.B., author
Title: EMPOWERED TO DECEIVE

ISBN: 978-0-6481255-2-5 (pbk)

Subjects: FICTION / War & Military
            FICTION / Family Saga see Sagas
            HISTORY / Europe / Germany

Also available as an ebook: 978-0-6481255-3-2 (ebk)

Typesetting and design by Publicious Book Publishing
Published in collaboration with Publicious Book Publishing
www.publicious.com.au

Dedication.

I dedicate the second book of my trilogy *Empowered to Deceive,* to four courageous women who inspired me to write: my mother, Edith Jean taught me to read, my second eldest sister Ethel [Julie] Wood encouraged me to be creative after she read the original manuscript. Colleen McColloch thought I had potential to write a novel, and put me in touch with Jean Easthope, a friend in Cessnock. Jean taught me to speed-read and she also inspired me to create my characters. Sadly, all these wonderful people are no longer with us.

1

Late in February 1946 Rolf von Breusch resettled in the small town of Sevelen, Switzerland. His hosts at the Edelweiss Pension were hospitable and considered his state of health. Registered under his Scottish name of Ian David Ross, they accepted him as an overseas businessman. Likewise, the financial institutions from Zurich through to Liechtenstein welcomed his patronage and funds, most of which were pilfered from his deceased wife's inheritance.

Since the war's end nine months earlier, Ian decided to reside permanently at this tavern. Contented and relaxed, he stayed there until the current influx of refugees from the Fatherland had either moved on or settled elsewhere.

Posing as a Scottish gentleman, Ian decided to wait for an opportune moment to mingle with the masses of displaced persons fleeing to either England or Spain. He knew it would be decidedly unwise to leave Switzerland.

Since moving to Sevelen he travelled to Zurich on a monthly basis. The city's banking facilities suited Ian as he pursued to ecru credit in the *Suisse* Bank and others in Zurich. This cunning ploy of diversifying funds gave the impression that money was constantly being invested from overseas dividends. Another subterfuge during his recent trip to Zurich was to arrange for substantial funds to be transferred to London and Ireland. If running short of money, he travelled to Liechtenstein where the banks housed substantial funds under both his assumed name of Ian David Ross and as a prosperous Irish landowner, Liam M Byrne.

After winter's thaw and on cloudless days Ian spent many pleasant hours roaming over the lower pastures or hiking through fields of spring flowers to the high plateau towering above the town. Time was his only master now. A private man, he preferred to leave his dismal past lost in the shadows of war, to live a cultural life in his self-enforced exile.

Embracing nature's warmth on a splendid spring morning he wandered down by the Rheine riverbank. Before ascending the foothills, he paused to appraise the spectacular mountain scenery outlined against a voluminous expanse of blue. Wisps of magenta shaded the Alps, frozen in a pillar of time.

Gathering a fistful of small flat stones, he skimmed each one across the millpond surface of the river. Wistfully he watched the pebbles bounce, leaving ever-widening ripples before sinking to the clear sandy bottom. In his childhood years this had been a favourite pastime while holidaying with his mother at Loch Levin, in Scotland.

Unexpectedly he gazed directly into a brilliant sun. Blinking, he thought his eyes were playing tricks. A shaded hand allowed his vision to grow accustomed to the morning glare. Silhouetted against the sunlit background he caught a glimpse of a shadowy form. It approached him from the bend in the river, which brought to mind one of his favourite ancient poems, *Camelot*. The petite figure swayed on unsteady feet. Unfortunately, the youngster collapsed before he could reach where she had fallen.

Lifting her, Ian tenderly placed her on the grass under a quivering aspen. Mystified how she'd managed to walk this far with her apparent injuries, he assumed she lived locally. Fascinated by the long eyelashes he bent over to admire her elegant features. Congealed blood on her temple indicated the cut must have originated from a previous fall. 'Her features are exquisitely beautiful. I wonder how she came to be out here alone.'

As a finger teased the fine strands from her forehead he noticed a graze on her right temple. Greenish-purple bruises were deepening on the slight swelling above the ear. A trickle of dried blood had concealed the wound on her pale cheeks. Unable to resist her lips, Ian stole a kiss as his finger roamed across their delicate skin, softer than silk to touch.

A sense of her awakening stirred his paternal instincts to life. Easing the girl up into his arms he observed her eyes flicker slightly as he thought: *She's only in her mid-teens. I'll take care carrying her down this uneven slope. By the look of those injuries she needs a doctor urgently.* He felt her pulse and found it threaded. Knowledgeable only to a degree in medical procedures, he detected a feeble beat.

Wera Traugott glanced at a dark figure approaching her, silhouetted against the strong sunlight. Sweeping the front path of the Edelweiss Pension, she heard Ian mumble something unintelligible. As she rushed to his aid the broom went flying and landed in a garden.

'Gigi what have you done now? Ian, carry her upstairs to her room. Number fifteen. I'll be there soon.'

Hesitant, he queried the room number. It sounded incorrect. 'A married couple's occupying that double suite, I think?'

'Of course, you're right. I meant room fourteen. I'll be there when I've spoken to Doctor Wilhelm. Stay with her, Ian.' Frau Traugott frowned, glancing down to where a guest was speaking on the desk telephone. 'Oh bother, I wish she'd hurry…'

Ian interjected, 'Yes, it's one room away from mine.' Surefooted, he carried the unconscious girl up the first flight of stairs. Forced to stand back to let guests pass, he again challenged Frau Traugott.

'This youngster's wound needed suturing on her right temple. I noticed her left knee is grazed and quite swollen. What about the door?'

'It's ajar. Imelda, the new maid has finished in there. Gigi's unlike you, Ian. She never locks her room every time she vacates it.'

'Business dictates I must, you understand?' On the top landing, Ian readjusted the precious bundle in his arms then stepped two paces to her bedsit. His elbow edged back the covers and he gently placed her in between clean sheets. Her sandals were discarded and the bedclothes pulled over her inert body. Feeling the clamminess of her skin warned him she may have gone into shock. Swiftly he tucked another blanket over the counterpane to keep her shivering body warm. Her eyes flickered again then closed.

Sitting by this drawn-cheeked youngster, Ian assumed her to be in her late teens. And as his hand embraced her wrist he let the graceful fingers rest on that palm. *She's incredibly beautiful. With her soft violet eyes, long eyelashes and cameo-pink skin she reminds me of a Dresden figurine.* He guessed by her smooth fingertips that they'd never done a day's work in their entire existence. Besotted with Gigi, he sighed. *How angelic she looks in her enforced slumber. She's the most exquisite creature I've ever seen. Dressed in finery I imagine her being a great lady, or a dancer of some renown.* The fingers of his free hand fondled the damp cheeks, now flushed by pain. His finger feather-caressed her lips, their ruby glow now faded.

Leaning on an elbow he stroked his goatee beard while appraising her unblemished skin. *Who is this girl? She's monopolising my thoughts. Is she an English rose? No, not with the French name of Gigi. She could be an entertainer, or perhaps a mademoiselle of wealth? Somehow, I don't think so. She is so unlike the French le femmes I seduced in my youth, before I began working at the Chancellery. She's totally different to the whores whom I encountered in my numerous official visits to sweet Paree.*

Ian's fantasising ceased when Frau Traugott and the doctor entered her room. He grunted, appraising them both with sullen eyes, full of mistrust.

'Leave please, Ian. You can return once I've examined my patient and dressed her wounds.' The doctor's glower warned Ian not to argue. Frau Traugott moved one step towards the door until he frowned. 'Stay Wera, I may need your assistance.'

The doctor's blunt attitude flawed Ian. Devastated, he grimaced. 'I'll wait outside.' The melancholy scowl on his ruddy-face deepened. It imaged a languid heart. Angry, he fumed. *I rescued her from a fate unknown, yet it wasn't good enough for me to stay.* Standing in the cold, draughty hallway he shivered. *What's more, I consider that charming youngster mine to care for. Not theirs to covet.*

This undesirable streak of possessiveness, indicative of his Nazi training, increased by the minute. The defiant trait well ingrained in his disposition caused Ian not to foresee her future, if she became entangled in his treacherous web of deceit. Intrigued by her shabby appearance, he now realised her torn clothes had occurred when she'd fallen on the moss-covered rocks.

Again his destiny took precedence over hers. These dictatorial rulings were similar to those which he'd imposed on his deceased wife, Erika. Their Jewish *kinder* remained but a faded memory in a world that no longer existed.

On several recent occasions Ian Ross had been confronted by his past in Zurich. One of the incidents he refused to acknowledge was unknowingly preordained. One evening he was accosted in a Swiss restaurant by a friend of his deceased father. Fortunately, the man had readily accepted his blatant denial of him not being the son of Rudolph von Breusch.

The other situation may have ended in a catastrophe. An acquaintance from his dismal past in the Reich, had addressed Ian by his Germanic title. The officer stated some facts that even he found repugnant. He implied that Captain von Breusch had rigged his own suicide before absconding to Switzerland. Under his Scottish nom de plume of Ian David Ross, he dismissed that officer's assumption as ridiculous, more so unjustified.

This information could have eventuated from one source only - the lieutenant who'd threatened him at the Swiss-German border. Suffice to say, the following morning that ex-Nazi officer was found lying in a filthy gutter with a bullet in his brain. With only *his* fingerprints on *his* pistol, the Police Inspector concluded, the weapon in their possession had been discharged accidently. An entry in his report read: "The officer's death must've occurred after the weapon was cleaned outside his hotel in Zurich."

In Sevelen at noon: About to escort the doctor downstairs, Frau Traugott gestured for Ian to remain by her injured guest's door.

'Doctor Wilhelm thinks it advisable for someone to be with Gigi. I suggested you Ian. It'll be for an hour or two, or until I can arrange for a live-in nurse.'

'My pleasure, Frau Traugott,' he agreed, with an exuberant smirk. 'What damage has she sustained, besides the obvious? No broken bones, I trust.'

'Fortunately, no. The doctor gave her a thorough examination. Delayed concussion, he said. The nasty gash near her eye he stitched.' The frau gasped in despair. 'Gigi's not to move off that bed for two days. I'll try to arrange for a temporary nurse now.' Wera Traugott listened for a cry. Nothing! 'Her knee has taken quite a jolt and Gigi must keep off that leg, or the swelling will not subside.'

Ian grimaced. 'I don't mind sitting by her bed. I realise you're busy, Frau Traugott. There are some personal things a man should not attempt to do...'

'No hassles there,' she interjected. 'Give me time to organise something. Gigi's on the cusp of sleep.' The frau listened. Looking at the room nearest them she nodded. 'You can go in now. Let her wake naturally, Ian. She needs to rest.'

'Enforced, so it appears! Who is this girl? A professional of some kind would be my guess, going by her soft, unblemished hands.'

'She'll be pleased … you calling her a girl. I know she doesn't look it. Gigi's in her mid-twenties. She worked as a professional dancer until a severe accident caused her intense pain.' A grin gradually appeared on the frau's wizened features. 'Until yesterday her manager was staying here. Pierre and Gigi have not long returned from touring through war-torn France. Long hours of entertaining Allied troops is the reason why she's so tired and listless.'

'Well, I never. She was a ballerina! Her slim physique and those sensual legs gave me that impression.'

'You men never miss a trick. Yes, she's a prima donna. Gigi has studied music and ballet all her life. She sang and danced with the English and Parisian ballet companies. Now sadly, she can't dance. Though she warbles with the sweetness of a nightingale.' A serried sigh petered through the landlady's wrinkled lips. 'Her pale Welsh complexion fools everyone. Its beauty mirrors her gentle nature and fine features. Not like mine, a shriveled juniper-berry!' She smiled as Ian's unshaven face distorted. The corners of his downturned, twisted mouth looked ungainly.

'I was right. Originally I assumed she might be of English extraction.' Pausing, his intense frown deepened. *Is this young lady a relative of Frau Traugott's? She speaks of her in such endearing terms. From memory, I haven't heard this charming creature mentioned by name. Well, not until an hour ago.*

Frustrated and feeling ostracised, Ian disapproved of the frau's determined pout as he queried, 'Frau Traugott, is she *your* niece? You seem quite fond of this young lady.'

'No Ian. I wish she were my daughter. We have no children. She and Pierre have stayed here on and off for years. I will miss Gigi, when she returns to Paris…'

Interjecting, again he demanded in a forthright tone. 'Tell me, is the Frenchman her lover, apart from being her manager?' A deep ingrained streak of jealousy shone through his gruff façade. Ian lent over the top banister as Frau Traugott moved down a step, to let several guests pass.

'Excuse me a moment. I must speak to the chambermaid who's working opposite Gigi's room. The least noise she makes will enable her to rest. I shan't be long, Ian.'

Impatiently his thumb strummed the wooden balustrade until her return. Aware he'd promised to sit with Gigi, he peered through the crack in her door. She hadn't aroused. From where Ian was standing in the dim hall he could see two silhouettes. He figured Frau Traugott was chastising an elderly guest. Gingerly he moved back towards the stairs.

The slightest knowledge of her background will be beneficial. And knowing what subject is close to her heart may encourage Gigi to confide in me. An endearing liaison with this enchanting creature might keep me from drifting back into the doldrums. Misery had served as his constant companion since absconding from Germany late in forty-four.

'Whatever gave you the absurd idea that Pierre and Gigi were lovers?' the frau demanded on her return. 'Of course, they're not. Pierre's a fine gentleman, one of the "old school" from Montmartre in Paris. They think the world of each other, both friendship and business-wise. Illness forced Gigi to forego her career in Milan and Paris. Pierre brought her here and I nursed her back to health. It's unwise to assume, or jump to conclusions Ian. And it *is* unbecoming of a Scottish gentleman.'

'I don't know, really. You gave me the impression they were lovers when you spoke about their close relationship. No other reason…other than idle curiosity.'

'Chef will be chasing my tail for tonight's menu, I must hurry. Call if you or Gigi need my assistance. I'll come straight up, Ian.' Leaving him to meditate, Frau Traugott walked in her stately manner down each stair. Near the kitchen she paused when accosted by an arrogant patron. A distinct wave of one hand dismissed the crank, who departed to catch her transporter. The driver, crunching gears, slowed to a stop. A notice read "This omnibus is out of service due to mechanical failure". Angrily, she plodded on with blistered feet, in worn shoes, the two miles to town.

Ian collected a disheveled army manual from his room then returned to Gigi's bedsit. He pondered how she had endeavoured to keep upright with wind whipping around her frail legs. *If I hadn't reached the riverbend in time, she could never have dragged her injured feet over rough ground. I doubt if she'd have lasted an hour in that bleak weather. Lying alone in the dark all night and chilled to the bone, she would have died of hypothermia long before dawn.*

Again his finger feather-touched the delicate flesh of her "perfect" lips and he grinned. 'Those drawn cheeks and pale skin look unblemished by lamplight. I wish she'd awaken or smile at me. This strange little creature owes me her life. I saved her from dying a terrible death in the wilderness where wild animals roam or hunt for their evening meals. What I find intriguing is why she ventured there under the hot, noonday sun.'

Ian read, while she slept. At one stage he imagined she stirred and lifted her fingers to his parched lips. Endowing the tips with a kiss, no movement of their own volition was evident. Quietly he cursed and spasmodically read to suppress his frustration. Fascinated to hear her life story, it goaded him to tickle her palm. However, he refrained from doing so when Gigi stirred. This gave him the impression she may have been foxing.

The little wretch, she's taunting me. Annoyed, he fumed. *Damn! I meant to ask Frau Traugott her surname.* In a deliberate ploy to awaken her, one finger persistently trolled across Gigi's cheek. Still, he failed to notice her eyes flicker. 'I suppose she'll tell me a little of her past life when she does awaken.'

He tossed the novel aside. It became a compelling desire to know more of this woman who had pilfered a corner of his heart. He couldn't resist the temptation to fondle her graceful hands. Holding those fingers to his lips, again he relished kissing their tips. Upon release, the unresisting hand fell limply on the counterpane.

This didn't faze or daunt him. The same persistent finger began to roam around her palm. As it edged up the lifeless arm, she flinched.

'Must you, Ian? I detest being tickled.' Gigi eased her arm away, tucking it back under the counterpane. 'I detested you touching my hands, especially when you persisted. It gave me the creeps, a feeling of crassness.'

'Oh, so you *are* awake? How long have you felt my finger creeping up your arm? I don't particularly appreciate being a puppet for you to dangle on a thread.'

'Just before...' Gigi paused. She held sway over this stranger. At a disadvantage, he looked at her with disdain. She enjoyed keeping Ian in suspense, 'I heard you and Frau Traugott discussing me...and what I do for a living. I also sensed jealousy in your voice when she mentioned Pierre. You needn't be jealous, Mister Err....'

'Did I sound that obvious? You *are* beautiful. I'd be a fool not to be fascinated with your beauty and your talent. Gigi is a pretty name. What might your surname be?'

Aware it could be dangerous to pursue the discussion, Ian resorted to another tactic. Bedazzled with this woman, he could make a fraudulent slip by disclosing his interest in the arts and his dark past. His pride escalated over being classed as a distinguished officer in Wehrmacht at the Chancellery in Berlin. And he found it profoundly gratifying with the position of honour he'd held in the Reich. Following this avenue of thought would amount to mortal suicide.

'Having not been formally introduced to you, it's absurd to keep referring to you by some fictitious name. Nor can I call you Mister Err… or whatever.'

Ian ignored her request. 'I suspected you were awake, as I carried you from the stream back to this inn. Don't bother to deny it, Gigi.'

'Well, maybe…a little I guess. I'll tell you the truth, when I know what to call you.' In a tantalising mood, she reveled in taunting him.

'Frau Traugott will take her broom to me, if I upset you again. It's…' He purposely faltered to teasingly prolong her agony.

Gradually her fingers sneaked above the counterpane. 'I won't tell you a thing. You cheated, Ian. Why must you torment me?' The smouldering look in her eyes conflicted with his steadfast gaze.

Still he hedged with giving her an answer. 'If you must know, it's Ian David…' A stroke of his well-established goatee concealed a taunting smile.

'Good, now we're finally getting somewhere, Mr David. It's an unusual surname? I've never heard it before now.'

'I didn't finish, because you rudely intruded. It's Ross,' he confessed with a smirk and a devilish look hidden deep in his insipid-blue, mysteriously curious eyes.

'Thank you for confiding in me, Mr Ian Ross.' A delighted smile impinged on her attractive features. 'I can't recall mine.' She hesitated. 'Gigi is my professional nom de plume. That's all you need to know…for now.'

'Who is Pierre? Truthfully, is he your lover?' Ian's shameful query caused her to blush, even pout. Determined to hear the answer from her lips he pressured her to reply. 'Well, I'm awaiting your response, Gigi.'

'You know damn well…he's not my lover. I find your inquisition offensive.' A harsh frown disfigured her youthful features.

Hell, have I blown this chance to be intimate with her? Regretfully, this isn't the way I wanted our relationship to begin. Accepting the folly he'd inadvertently made, Ian decided to call her bluff. He gave an apology the only way he knew how. Sorry, please or thank you didn't exist in his vocabulary.

'Can we start again? I certainly didn't mean to offend you, my dear.'

'Didn't you?' she snapped. 'Frau Traugott told you Pierre is a friend of ours. Don't you listen, Ian? Forget it. The subject is now closed. I don't wish to discuss it again. Otherwise, we will never be friends.'

Try as she might the severe pain in her leg and head restricted her movements. *Is this man insensitive to my stress?* Sulking, she refused to talk until he showed a little compassion for her. In agony she again grimaced.

'Allow me! There you can move now Gigi.' The sheet and covers were lifted off her body. Not prepared for this sudden onslaught, she wasn't quick enough to anchor the brief nightdress around her naked torso.

'Ian, *please!*' Embarrassed, she reefed the bedclothes back and tucked them firmly around her shivering shoulders. Even though the sheet hurt her bruised knee and hip, she swore he would not get another glimpse of her nakedness.

'My dear, I have seen a nude woman before. Besides, I looked away and didn't see your bruised breasts or legs.'

She knew he was lying. He'd given himself away by this declaration. Normally she didn't wear night attire. *Frau Traugott insisted I must wear a nightdress when she told me the doctor was on his way. No doubt, like most men, Ian relished seeing me nude.*

'One day Gigi, you and I will have a good laugh over this incident. Then I'll take great pleasure in reminding you of your immature modesty.'

'You're a cocky Irishman! If that's the brogue I detected, Ian?' She then thought, *No, his diction sounded Scottish. There's something amiss with his accent. If I keep taunting him, this fibber may reveal the truth of his nationality.*

'I don't believe you Ian. You're not an Irishman.'

'I've lost most of my accent. That goes from living long years away from Inverness. My mother wouldn't recognise me now. Not with this goatee beard and a well-trimmed moustache, although I did go home to visit her last summer.'

Ah, so he is Scottish. Now that makes sense. 'Your mother would be quite elderly now, I suppose? Wish my folks were alive! I have no one other than...'

'I'm not buying into that argument again. Yes, she is. Though she's seventy-five, give or take a year, and not in the best of health. My father died two years ago.'

A weary sigh petered from her lips. 'I wouldn't have survived the war, if it weren't for my friends. Now, it looks as though another must be added to my Christmas list. Only if a certain person behaves himself, Ian.'

'Does that mean…I'll be noted on your list of special friends?' Need she answer? By the glower he received, it was obvious.

Smiling, he moved closer to the bed. Clasping her hand, he endowed her clenched knuckles with an endearing kiss. She pulled her hand away and so did he, as if offended. Exasperated, she frowned. 'Please don't annoy me. Provided you don't kiss my hands, or show the green eyes of a jealous dragon again, you *will* be on my list. And no more impudence or peeping at me. Do you realise that you made me blush? I hope you're aware of the embarrassment you caused me, Ian? You saw more of me than you let on.'

'Who'd ever believe I would tell a blatant lie. Not me.' A trained Nazi tactician in the field of deceit and a master at concocting fantasies, Ian used the persuasive power of coercion to make her believe he *was* genuine and not lying. He sensed, by what she had implied, that he was well on the way to regaining her confidence.

Comfortable in the knowledge that he wouldn't hurt her, Gigi began to doze. The injection the doctor had given her was having an effect. Somewhere in her bewildered mind, she sensed Ian might be spying on her through the wardrobe mirror.

Instead, he stood passively gazing through the long bay windows open to the sky. Pensively he focused on a snow-capped mountain towering high above the distant border of Austria, and a legion of queries strafed his mind regarding this mysterious woman who haunted him. *Why she refused to reveal her surname seems ludicrous to me. It's as if she's concealing something devious of her past life, beneath that delicate façade she poses to the world.*

Frau Traugott ventured upstairs to relieve Ian around four. Assisting Gigi with her personal toiletries, she helped her to prepare for the night. Ian had heard Frau Traugott mention that her doctor intended to drop by in the morning to assess his patient's knee and to rebandage her forehead. The sutures would be removed within six to eight days.

Around five-thirty he carried up her tray, then returned downstairs to collect his meal. The instant they'd finished eating he insisted on reading to her. Within half an hour Ian had worn thin his welcome. Gigi had tolerated his wearisome attentions long enough. In between the persistent pleasantries he kept fondling her fingers.

Eventually on the verge of screaming and desperate to retain her privacy Gigi ordered him to leave. 'If you don't return to your own room now Ian, you'll force me to use the bell Frau Traugott left on that dressing table.'

Determined to be within call, he refused to leave, until the novel he'd been reading collided with his head. 'I resent your pugnacious attitude and your rudeness, Gigi. That damn book just missed my glasses. All right I'm going. Don't think I won't retaliate or let you dismiss me in future. If you throw anything at me again, you *will* be sorry,' he snarled and thundered from her room, slamming the door behind him.

Next morning 7am: Ian repeated his charismatic endearments which annoyed Gigi until she almost screamed. His jealousy festered until the cauldron of mistrust began to bubble. Amidst constant snide remarks he kept mentioning Pierre Jean Paul Bouvier until she found his persistence intolerable. Ian's frequent petting she abhorred. She detested his overbearing nature. Her only recourse was to spurn his advances.

On the third day, she found sitting in the chair more comfortable. The swelling in her right knee had subsided, enough to retain her balance. A slight limp was evident when she walked, or if she placed her full weight on that ankle.

After dinner Frau Traugott paid her special guest a courtesy visit. For an extensive time, they discussed Ian's spurious and overburdening manner.

'Frau Traugott, I can't stay here another week. Ian's flirtatious innuendos and his continuous fondling of my hands are driving me insane. Can you think of something to keep him busy...or away from me? Whatever happens, I'm determined not to return home to Wales until I'm well enough to face Manny. Poor dear, she would take a fit if she saw me in such an anxious state. Look, my hands are shaking. If he doesn't stop pestering and patronising me, I'll leave tomorrow. His persistent groping of my arms I find intolerable.'

Manny, her elderly grandmother would be overcome with shock if Gigi arrived home distressed after her recent accident. Tired and feeling languorous, she dreaded leaving the Edelweiss Pension and Frau Traugott. The elderly innkeeper considered her more of a daughter than a guest. The elderly landlady treated Gigi with loving-kindness, and had nursed her through this and many an illness.

Another bane of her dysfunctional life was how this cocksure Scotsman had drawn her under his charismatic spell. Each day Gigi felt powerless to break their friendship. This overindulgent and dominant man was manipulating her every move and restricted her freedom. Some of his attentive mannerisms she found irresistible, although she was incapable of repelling his advances.

Alone in her room and feeling miserable she wailed pitifully. 'Why am I attracted to this man who keeps intruding on my privacy? If Frau Traugott doesn't keep him away from me, he'll force me to leave.' Exhausted she tossed her pillow on the floor. 'Please God, make Ian realise how he's damaging my health. I don't know where to turn without *your* help and guidance.' The same prayer she repeated kneeling by her bed, before retiring at night.

Earlier in the week Ian had told Frau Traugott and Gigi that he had once tutored young pupils in a private Glaswegian school. They both suspected his ramblings concealed a devious ploy to hide a disrespectful or indelicate incident during his adolescent years.

If he worked as a teacher it means nothing to me. Money seems no object by the way he's spoken. I know he's told me ultimate lies, Gigi mused, listening to Frau Traugott explaining how she intended to keep Ian occupied. Hearing him ascend the stairs Gigi continued to think. *Who's to say he isn't fabricating more lies, for his own gratification? After his insincerity to me, how can I trust him? Ian's a devious man who'd concoct a tale of woe just to keep me under his thumb.* Confliction with his outlandish ideas still running rampant through her confused mind, she needed time to think.

By nightfall she felt really ill and oppressed by his obsessive attention. 'Tomorrow, I'll take you for a walk down by the stream. It'll strengthen your leg and we can converse on the way. I'm interested to hear of your life in Wales, my dear.'

'No Ian. You know dash-well I couldn't walk that far. Now, will you please go and let me rest. Don't even consider venturing back here in the morning.'

Unsettled by his undignified behaviour she insisted on him leaving that instant. Desperate to keep him at a distance, she acknowledged to herself how to prevent him from plaguing her to the extent that he was driving her crazy. This one-sided relationship was overpowering. She also found his constant endearments extremely annoying.

Early on the fourth morning Ian Ross sat impatiently drumming his fingers on the dining room table for her to join him for breakfast. To fill in time, he read the first edition gazette. After an hour he decided to find out why she hadn't arrived downstairs. *Could she be ill again?* Worried, he hurriedly returned upstairs.

Frau Traugott, in the midst of attending to another guest's breakfast, frowned at Ian as he passed the kitchen door, then continued speaking to the chef.

On the first floor Ian Ross rapped twice on Gigi's door. No response. He tried the doorknob again. It wouldn't yield to pressure from his sweaty hand. Silent and in a morose mood he stood to think for a couple of seconds. Perplexed he swore, muttered something and hurried down to the reception desk.

'Frau Traugott,' he called again. 'Damn, where could that woman be?' There was neither an answer nor a sign of the landlady anywhere. A brief search outside the tavern proved fruitless. In a dictatorial stance, he stood with thumbs tucked in the hip pockets of his grey trousers. A cool breeze ruffled several strands of Ian's dyed mid-brown hair on a clammy forehead. His furrowed brow oozed sweat and trickled down flushed cheeks. He scowled and knocked on her office door. 'Where the hell is that damn woman? I need answers now, not tomorrow.'

2

Bidding farewell to some of her guests beside their car, Frau Traugott turned to see him standing behind her. 'What is it, Ian? Going by that frown, you're not in a pleasant mood this morning. You seem agitated...'

'Where is she?' came his rude intrusion. When Wera tried to speak, Ian glared at her, and raised his voice dramatically. 'I asked you...Frau Traugott, where *is* Gigi? It is imperative that I speak to her...*now.*'

'Calm down and follow me through to the office. I refuse to discuss personal details of her privacy here in public. Ian, where she is should not be your concern.'

'Why not? You know how fond I am of that young lady, whom I find enchanting.'

His landlady scowled. 'I said we would talk inside, or not at all.' Annoyed by his irrational behaviour, Frau Traugott stormed off to the seclusion of her office. On entering it, she kicked the door closed with a foot.

'Sit down and don't interrupt me.' Settled behind the desk, she gave him a sombre glare, followed by a harsh rebuke. 'Right, you wanted to know, so I'll come straight to the point. You deliberately tried to force your attentions on her.' Warned by a threatening finger, Ian remained silent. 'I asked you to take care of her. Not hound or suffocate her. Gigi found it impossible to remain here with you stifling her every move. You were stupid to keep hounding the girl by persisting she cower to your demands, Ian.'

Nobody had ever referred to him in such a tawdry or trite manner. *How dare this hausfrau, a mere peasant, think of doing so to me?* An acrimonious glower displayed his displeasure. It equalled hers. This ex-Nazi officer recoiled in the chair. He knew there was little chance of reconciling with the woman he couldn't dismiss from his mind.

'Frau Traugott, I meant her no harm. My intentions were and are honourable. I think she is a charming youngster. I have studied many people and their lifestyles in my job as an overseas journalist, especially in Holland…and in Bavaria. Why you seem to think I would intentionally harm Gigi is ludicrous.'

Swallowing, words stifled within a restricted throat. Ian scoffed at the idea of this woman interfering in his private life. He realised she couldn't be fooled much longer.

'Ian, far be it for me to interfere in your private affairs. I'm not a fool. I am aware that you care for her deeply. Don't deny it, or mislead Gigi.' Frau Traugott's comments came as a warning, not a threat. 'You must be sure of your feelings for her. She's inexperienced in the ways of love. It would be inadvisable of you to hurt her in any way. Because she's an impressionable young lady, I'd hate to hear of her being abused by *you,* or any man whom she considers honourable.'

Perhaps my loving her sounds ridiculous to this witch of a woman. Still, I can't tell her that I have fallen in love with Gigi. Desolate, he sighed then responded, 'Tell me Frau Traugott, will she be coming back here, or has she returned to Llangollen? It's essential that I talk with her today.' These words puffed out in an unsteady stream. Reassured by his motivation and adoration for Gigi, it allowed him to breathe easier.

'Prior to her departure with two friends last night, she left this note on my desk. Read it, before commenting further.' It sailed across the desk and landed in his lap.

Quiet and reticent of mood he remained seated and ingested the contents several times. Unable to believe his eyes Ian stood; flipping the note in the air he trounced to the window. Leaning against its sill he peered at the rising mist and an icy chill travelled down his spine. This peculiar feeling was bleaker than the snow-saddled Southern Alps. Her letter bordered on an ultimatum.

How can this be? This implies that she will either return to Wales, or stay for the remaining four weeks in Appenzell with friends. If I want to continue our friendship, I must do so without making her feel persecuted. Ian sighed. *Well, I'm relieved to know that she'll be in here for breakfast, a week from today.*

'Frau Traugott, do you think it possible...' Stunned, he couldn't think what to say. Misery swamped his judgement. She waited patiently. In desperation he declared, 'Would it be possible for me to meet her on neutral territory in a café, perhaps around eleven tomorrow morning in town?'

'No Ian, it will not be convenient. I know you mean well. As a friend, I'm warning you not to proceed along those lines. You have offended her by constantly bringing up Pierre Bouvier, with the strong inference that he was her lover.' Allowing time for the rebuff to sink in Frau Traugott remained silent for a minute or so. 'Accept my advice. Give the problem more thought. A lot of soul-seeking before you wreck your future together. She thinks highly of you, so don't abandon hope. Gigi made me promise not to reveal where she is staying. I think she made that point perfectly clear in the note.' It now took pride of place over her acier-streaked head.

'How can I atone for my blunder? Flowers, chocolates! Which would she prefer?'

Frau Traugott sat back, her mouth agape. 'I don't believe this. I just told you to leave well enough alone until Friday. You don't listen, Ian. You'll cause a huge rift in your friendship with her if you persist with this nonsense. Sit tight and shut your mouth. You were blunt with me. Now I'm returning the favour.'

He responded with an exasperated gasp. 'It looks as though I've no alternative than to bide my time. It's going to be a hell of a week. Should Gigi ask how I've accepted the news, convey how devastated I feel. It's unbelievable to think she's offended to this extent by my innocent platitudes. I'm concerned about her health, I mean her no distress.'

'Tell her yourself on Friday. Now I must go, Ian. We expect more guests and Zelda, our kitchen maid is sick. The scullery maid is working in her stead.' As Frau Traugott stood to usher him from her office, she tapped Ian on the left shoulder. 'To save you moping in your upstairs room, you can work in the kitchen. You enjoy cooking?' She accepted his nod as yes. 'You'll assist the chef. Hard work will occupy your mind, instead of dwelling on her departure. Make sure it's not a permanent rift, Ian.'

'In my youth, I often enjoyed searing venison and tender loins of veal at home, or preparing family meals. It's belittling to do scullery work. I'm not adapted to coping with such menial tasks.'

As an ex-Reich captain, Ian couldn't imagine himself doing mundane work. Still! *If it puts me in a favourable light with this old duck, I might try. Gigi is my boarding pass to freedom and I think she does care for me, a little. With the frau on side, she may put in a good word regarding my genuine feelings.* Rethinking over the prospect, Ian reconsidered the idea of soiling his hands by doing manual work with manicured fingernails.

'Frau Traugott, I'll agree to assist your chef. How long will he need my services?'

'At the moment several days, could be longer. We'll expect you in the kitchen an hour before each meal. I'll dock your wages from the weekly rent. That's if you're not offended, Ian?'

A thumb gesture indicated it was fine, yet he bordered on rescinding her offer.

'Good. It'll alleviate a lot of strain from our workload and I will be grateful. I may even make it my business to speak to Gigi on your behalf, when she phones.'

Tempted to say, give her my love, again he realised this was none of her business. Then he reconsidered his plight. *It might be advantageous and may work in my favour, also it could improve our relationship.*

Frau Traugott told Ian to be in the kitchen by eleven. The broadest of smiles softened his demoralising grimace. 'I can't have my favourite guest lost where duty…' The front desk bell tinkled. Scowling, she left Ian to ingest their talk regarding his control over this Welsh woman with whom he was besotted.

Seven days seemed an endless drudge for this lost lover. Thinking of his life in Germany, he pondered, *Gigi's scintillating wit and gracious smile far outweighs my deceased wife's beauty. Her poise equals Madeleine's Italian mannequin-style of elegance. I learned so much about sex from her. Heide lacked composure. Gigi's charisma outshines Maddie's. The Parisian prostitutes I associated with in Chatillon were accommodating. I've yet to meet a woman who can compare with Gigi's looks, figure and brains. My fascination for the intellectual power she holds over me can never be dismissed lightly. She is the ideal woman, all men desire in a loving relationship.*

Ian felt as if he were being penalised for his kindness, more than his misgivings. Long walks or hikes he normally enjoyed were reduced to brief meanderings. Her denial of him dragged his moods down in the muddy waters of despair. Rain weeping across Sevelen was of little or no consolation. Unscheduled winter winds heralded untimely coal, or wood-burning fires.

With an hour to spare, Ian sat in the solarium to drink his coffee. With his mind cocooned in the clouds of yesteryear, he failed to hear someone refer to him by name.

The broom Herr Traugott wielded whizzed past his knee. 'Ian, my old frau's been looking for you. You'll find Wera at the front desk.'

'Yeah, I'll go straight in there. She probably has another rotten job lined up for me.' Entering the vestibule in a hurry, he almost collided with Frau Traugott.

'We'll speak in my office. Not here Ian.' No further words exited her staid lips. Not until they were alone and the door closed. She noticed his down-trodden look. His moroseness didn't appeal to her. *I'm tired of living with that glum look. It really does mar his handsome features.*

'What's wrong now, Frau Traugott? Has *that* wretched Frenchman contacted her again? Or is Gigi ill?' Ian stood in an officer's stance with both hands clasping the butt of his work-scrubbed, dull-grey tweed trousers.

'Not exactly,' she admitted. 'Sit down and listen. I have received a message from her. It regards your disagreeable behaviour, Ian.'

'Don't hedge. It sounds as though she's still annoyed with me.'

A harsh glower cautioned him to desist ranting to let her finish reading. 'She's quite well...' the frau hesitated, her attention drawn to childish chatter of infants romping in the corridor. 'It appears Gigi will be here...around ten this morning.'

A short sharp gasp signified Ian Ross must remain silent. His mind travelled along a different avenue. *This is terrific news. She may not spurn my advances again. It might be prudent if I keep my mouth shut and listen to her excuse for leaving without saying auf Wiedersehen.* He pondered, *I may've been a bit hasty in my appraisal of Gigi. If we're on speaking terms, I won't risk offending her a second time. Trying to cope alone this week has been sheer hell for me.*

Frau Traugott intruded on his thoughts. Calmly she remarked, 'You really have taken a shine to Gigi.' The frau paused until Ian stopped sighing. 'I'm pleased to say, she's feeling more contented after her brief absence. She'll be home within the hour.'

Happier than a reprieved criminal, he stood on aching feet. Nodding to Frau Traugott he declared, 'I can't believe this ordeal is over.' Ian sighed. 'Not to have seen her for another week, the suspense would've driven me crazy.' Increasing his dulcet utter, he spoke in normal tone. 'I feel relieved now my anguish is over.'

'Think again, Ian. No, it is not. One false step on your part this coming week will be enough to send her packing. Gigi states in this note that she intends to fly back to Wales. I'll book her flight. She'll strike no hassles, because of the concerts they both did to comfort our troops in the field of war. Entertainers are never left behind when space is available.' Sorting the mail while talking, Wera Traugott looked across at her disillusioned tenant. 'My advice is for you to tread warily. Caution will be your ally. Take each hour slowly. Don't rush or say anything that you do not mean which you may later regret, Ian.'

'Well, it appears I must be patient.' A flexed shoulder conveyed his plight.

'Now, that is a sensible approach. I won't need your services from noon today.'

Comfortable with his decision, Wera deducted two weeks rent from her records. With the money saved, she advised Ian to invite Gigi to dine under candlelight in a quiet eatery in town. 'She'll enjoy your *pleasant* company in the new restaurant with outdoor dining facilities.' The frau scrunched *her* note she'd read, in case she missed something.

Consoled in the idea of *her* possible return, he chose not to respond.

'If what I've observed is correct, Gigi has a special spot in her heart. Yes Ian…for you. Don't ruin your final chance. It might be prudent of me to admit, my staff and I may be catering for an engagement party, before she leaves in three weeks.'

Shocked, a gasp expressed his disappointment. *Three weeks! Gigi told me it would be later in May. I didn't expect this news. The slyness in those impenetrable eyes of Frau Traugott says she's aware of my sudden reaction. It means I'll be stuck here in Sevelen while Gigi's enjoying that Frenchman's company, in France or across the Channel.* With a deflated ego and a heavy heart Ian walked to the office door, rested one hand on its brass knob and turned as Frau Traugott spoke.

'Gigi is trying to arrange, through the Chargé d'affaires here, transport for you both to London. Mr Homer Ellis is a close friend…of hers.' Frau Traugott refrained from saying Pierre in fear of stirring to life a fresh argument. 'If you agree and things work out between you, Mr. Ellis will organise your flight tickets.' She pointed to Ian's gaping mouth and warned him to keep it closed. Diplomatically she suggested, rather than to antagonise him, or inflame a difficult situation. 'I think you should go upstairs and make yourself presentable. Wear your grey suit and a demure shirt. Perhaps, the pale green one and your paisley tie with soft cinnamon overtones. Gigi admired those clothes the first night you took her to dinner in town. When you come downstairs go to the recreation room, in there neither of you will be disturbed.'

Frau Traugott tossed Gigi's telegram in the bin, then walked to the filing cabinet. Turning she gasped. The lovebird had flown.

Over the ensuing four weeks this engaged couple discussed whether Gigi should resume her career on their return to England. Opposing this idea, which he considered ridiculous, Ian disputed her plans. In a fuming rage, he repudiated her decision and suggested it should be delayed until they arrived in London.

On an odd occasion Gigi had asked a lot of unanswerable questions regarding his private life in Scotland. Her inquisitiveness grew daily. Though little did she realise this was an unhealthy topic to pursue. A foolish comment or indiscretion of Ian's could prove fatal to their relationship and to his life.

3

A week rolled on and the couple were almost inseparable. The fresh morning incited Ian to request her to go for a walk. Nothing strenuous was suggested. Her sprained ankle had progressed as expected and another accident now could be disastrous.

Hand-in-hand they talked while rambling down a lane towards the Rheine canal. Having promised to describe her younger years in Llangollen, she now began to regret this decision. A shade hesitant, Gigi responded solemnly, 'Ian, I don't wish to bore you with my uninteresting life there, or in Paris. Although I would like to know what you do for a living.'

'Now, you mean?' Ian saw her nod and followed through by saying, 'I conduct my business affairs from here, in Sevelen. As you must be aware by now, I am jealous of two things - you ogling other men and my privacy. Gigi, my plans will alter in London and I'll find a cottage with a garden to keep you busy. It's as simple as that, my dear.'

'Sounds interesting and logical,' she nodded casually kicking a patch of deep blue cornflowers with the toe of her cream sandal. 'My ideals are different. Because I only have eyes for you my darling. What other interests or hobbies do you enjoy, Ian?'

'I sometimes teach private pupils in Zurich, although it's been monotonous work until now. With you by my side, our marriage will be ideal.'

'You're a strange man. What a peculiar proposal,' she laughed. 'A girl ought to make you wait for an answer. Will it be yes, perhaps? Or should I decline your…'

Ian ceased walking and grasping her, he swung her around to face him. 'Gigi, you will *not* abandon me a second time. Your non-reply I'll accept as *yes*.' Edgily he was forced to admit, 'I love you beyond all reason.' Looking skywards, he confirmed, 'Now we're in for it. Let's run to the ski hut, or we'll be drenched with rain.' Spots larger than golf balls descended from angry bluish-green formations swirling above their heads. Thunder rumbled, lightning zigzagged across the obsidian sky. The heavy cloudburst expelled huge clusters of hail. Wind gathered speed which engulfed everything in its path. Violence reigned.

Unable to keep pace with his quick stride, her hand slipped from his grasp. 'Ian, I can't hurry. Help me across this uneven patch. It looks unsafe and I could trip or fall.'

Sweeping her into his arms, he carried her over the sodden earth until they reached the hut, ten metres away.

The previous evening her rebuttal of his amorous flirtations had caused a colossal argument. Fearful of losing her virginity she'd hesitated. His persistent heavy petting she found intolerable. His burning desire to seduce her was insufferable.

On a sudden impulse she surrendered to his charms. Her resistance melted when she'd heard the words, "I love you, beyond all reason." Where had her inept sense of reasoning flown…higher than the clouds, now sprouting a torrent?

Soaked to the skin on entering the ski hut, he endeavoured to ignite the dry kindling stacked in piles for emergencies. Looking at her, Ian ordered her to strip naked 'Gigi, remove those wet clothes while I persevere with this damn fire. Once the flames burn blue, my arms will keep you warm. We can sip our wine together on a mat in front of the fire.' Two goblets sat on an uncluttered bench and he wrestled with his battered hip-flask until he eased it free of his damp jodhpur pocket.

Scouting around he retrieved clean blankets from a cupboard. Fresh towels from a shelf sufficed to wrap around their shivering bodies. Drying her long hair by the hearth her arms embraced the wooden mantelpiece to warm her breasts and freezing body. Her legs felt numb in the chill of day.

When the fire was alive with flames Ian stripped down to his under-clothes. The wet trousers, shirt and underpants were spread eagle-fashion over the legs of an upturned chair. This maneuver rekindled the image of the first night he'd spent in Potsdam after his planned ensconce from Berlin.

With a strong blaze underway Gigi began to disrobe. Partly naked, her bashful smile encouraged him to unclasp her ecru cotton bra. Ready for love, Ian reveled in the idea of removing her cream, coffee lace-edged panties, which he tossed over a wooden stool.

Cradling her breasts from behind, he imagined them to be smaller than they were. Tenderly massaging her nipples, he felt the soft flesh gather and pleat beneath his fingers. In a moment of salacious bliss, he encouraged her to free the towel tucked around his waist. Gently maneuvering her body around to face his, Gigi could feel his firm probe pressing hard against her pubic hairline.

Ian was the only man who had succeeded in seducing her. Not even Pierre, with his gentlemanly French manners of persuasion, had seen or fondled her naked breasts.

A bit hesitant to permit Ian the freedom to fondle her body she gradually relaxed. Slowly he began to induct this sweet creature with the gentle art of foreplay. Massaging her thighs, it stimulated and excited her until Gigi allowed him access to her virginity.

He portrayed the part of an amorous lover, by plying her with sexual fantasies that she never imagined possible. She responded to his seductive moves with gracious oohs and aahs. Feeling secure in his arms, she drifted off to a contented slumber.

Upon wakening an hour later, she sensed he wanted to seduce her again. With the day still in its infancy, there would be ample time to pursue this pleasure.

Entwined in nakedness with only their shallow breathing to contest the tempest's roar, Ian meditated on their union. *I enjoyed the tender way she submitted to my charms. Now I've conquered her fears on promiscuous sex, Gigi, at my command, shall allow me to seduce her at will. She fought a valiant fight to retain her virginity and lost.*

He shivered as an ice-borne wind slithered between the hut's weathered timbers. The whistle of Nature's roar also made her cool body quiver. Her shudder rebounded on his nakedness. Edging the blankets around their bodies he fondled her curvaceous breasts. In a reflective mood, he pondered what their future might hold. The conquest of challenging her resistance failed to quell his instinctive desire to copulate again. In Germany, France and in Switzerland the women he'd seduced couldn't compare with this nymph's seductive endearments. Gigi more than satisfied his lust, she climaxed on cue and excelled in the art of physical love.

While she dozed he thought, *why should I enlighten this woman in my arms about my past romantic liaisons. Records of my marriage and kinder were blown to smithereens the day Hamburg was annihilated by Allied forces in forty-three.* Why, without necessity, would any man awaken a sleeping giant by mentioning his personal history and sudden ensconce from the Reich? His previous sexual conquests belonged to a distant era where murder reigned.

Should something be disclosed of his marriage, he intended to declare: *The sudden loss of my wife, forbade me from divulging her death. I'm the same as all widowers; lost and looking for love with a woman who could quell my sexual desires and needs. Why should I now confess to my liaisons with the 'ladies of the night' to Gigi? It's none of her business. She'd take flight if I revealed my dubious life in Berlin.*

Bending he gave her ebony hair, still slightly damp, a kiss. Nestling into the flesh of her curvaceous breasts, his lips caressed the tight-budded nipples.

In front of the fire Ian made a silent vow. *The sound of Mrs Ross has a magical ring. Our wedding will be a small affair, when and where I will decide. Not her.* Snatching another stolen kiss, he gasped with delight. 'You, my darling will allow no man to embrace your fragile body with his intent to seduce you. From this moment on you are mine alone to adore, to cherish for all eternity.' *Thanks to the powers above, no kinder will spoil our utopia. Those brats of mine were fried to cinders long ago, with their bitch of a nanny, Heide. They're both stoking the fires of hell, with their mother. It's quite gratifying to think, none of them will again threaten me, or my future with Gigi.*

Awaking she realised dusk was closing in, and yawning, she suggested, 'Darling, we should dress, as night will soon be upon us. My swollen ankle's still tender. How can I hobble back to your car in the dark?'

'Gigi, the rain's easing so let's make a dash for it. I'll help you. Hold my arm then you won't slip in this thick sludge.' In between downpours Ian managed to drive through quagmires of mud, diverted around mini-landslips until he pulled in beside the Traugott's vehicle. The duco on its rickety doors were rusted through in patches.

In the Edelweiss foyer, Gigi kicked off her ruined sandals. Ian hurried ahead to the kitchen to ask Frau Traugott about their evening meal. Around nine, and in blissful mood, and at his insistence, they showered together and then slept in their own rooms.

A raging storm shredded the leaves of aspens. Strong winds sent broken boughs flying against her bedroom windows until Gigi was forced to huddle under the blankets. Tempted to approach Ian's room, she shuddered as a blinding flash illuminated her room. A deafening rumble followed. Suddenly, peace and silence crowned the pitch of night.

'What a relief. The storm's passing. Now I can breathe.' Peeping above the covers she took two deep breaths while enjoying the cool, pink-lucilla and garcinia scented air.

Around nine the following evening she retired. Exhausted, after helping Frau Traugott in the kitchen all day, Gigi found it an effort to climb the stairs. She showered, wrapped a warm robe around her cool shoulder then dived under the goose-feathered eiderdown. Within a few minutes Ian again intruded on her privacy. His motive – to make her promise not to mention their engagement to Frau Traugott or her crank of a husband.

'Sorry daring, you're a little late. She knows about our wonderful news…'

'Why the hell didn't you consult me first, before confiding in her?'

His bleak mood and harsh scowl frightened Gigi. In retaliation she fumed. 'Ian, I think *you* are an *inconsiderate, egotistical* man. I detest you bullying me. Go back to your room. Leave mine now. Sleep is important to me. My ankles and feet are swollen…'

'More fool you. I'll leave when I'm ready. Not before.' Reluctantly he accepted her ungracious snarl. 'In future, keep your mouth shut and let me make important decisions.'

Ian hadn't the vaguest idea, nor would approve of what their hosts had proposed for the following night. Standing by her bed, his deep voice held a threatening connotation along with his distorted smirk and savage overtone. 'You will obey me, Gigi. I have an appointment in Zurich tomorrow at nine. And you *will* be coming with me. We'll have lunch together in a small, out-of-the-way café. Then I'll take you on a tour of the city. Time will be ours to forage through boutiques or wander along the gold-embossed plazas without people asking dubious, unanswerable questions or disrupting our privacy.'

'Ian, I can't go to Zurich tomorrow. I shouldn't be telling you this. Frau Traugott has arranged a small soirée especially in our honour. If you're not home by seven, Wera will be hurt. She asked me to help with the cooking, and I can't disappoint her. Before you leave, can we have a photo taken together under trees in the rose garden...'

In an outrageous temper, Ian raised his fist and objected. 'How many times must I tell you. I'll agree to attend the party...if no photographs are taken of me. You seldom listen to what I say. Why, Gigi?' Slightly bending, his lips brushed against her forehead. His left hand slid beneath the quilt to fondle her breasts. To be accretive he squeezed her right nipple.

'Ouch, that hurt. Why must you be so cruel, Ian?' Forcing his fingers away from her right breast and her throat, as a loud shriek pierced the night air. 'A cramp in my left calf has knotted. The lump under that knee will go once I relax.'

'One drink with just the Traugott's must suffice. A soirée is ridiculous. Count me out.' He pretended to leave until she held the wandering hand against her flushed cheek.

She responded to his glare with a despondent glower. Her sad eyes pleaded for him to rescind both his decision not to attend the party and his hasty departure.

'Darling, please don't let me down. I won't be able to face the couple if you refuse their generous offer. Have you forgotten, Frau Traugott convinced me to see you again? This morning I iced a chocolate cake with lemon swirls. It's ready for our special night.'

I must be a fool for letting her talk me into attending their stupid soirée. I refuse to let Gigi force me to do something that she will later regret. In all probability I won't be home until nine, if then. This unspoken snarl passed through curled lips; yet he yearned to love her physically. His previous comments reeked of bitterness. Reconsidering their discussion, she ignored it and his arrogance.

'Overnight you might rethink that harsh decision, Ian. When we cut our engagement cake, will you let one of the guests take...' She pouted as he frowned.

'I said *no* photographs, I detest having mine taken. I won't warn you again.'

Embittered by his discourteous rebuttal of her idea, Gigi hobbled across the room. 'Please listen to me,' she whispered, placing a finger on his chin. 'A photo of us together will be a reminder of how happy we are, or have been since you proposed to me. Later we'll reminisce and look back on our engagement night with...'

'I said twice no photographs,' standing by the door he objected strongly, 'and I meant exactly that.' Rudely he thrust her aside. 'Please yourself, if you want one taken on your own, I won't object. I refuse to have mine take by you...or some other fool.'

'A fool! How dare you call me a fool?' Catching him unawares her undamaged foot collided with his thigh. The force behind her anger sent him flying. He landed in the corridor. Mumbling about his despicable attitude, Gigi slammed the door in his face.

Enraged, he responded by lashing out at the architrave. 'None of my subordinates in Berlin would address me in such a tawdry manner, or treat me so abominably. I won't tolerate the bitch's rudeness. When she least expects it, I'll put her over my knee. A flayed whip, or sizzling-hot poker on the spine or feet does wonders. Nails removed with pliers worked well to force saboteurs and criminals to confess their crimes.'

In a rancorous mood, Ian flung the door open. The hard thump on her dressing table with a closed fist made her jump with fright. Spewing fire, he approached her with a blood-curdling whine and hate surfacing in angry eyes. 'I'll compromise, provided you never slam any door again. I would never have envisaged you being nasty to that extent. No photographs, I'm camera shy. You should know that by now.' A cuddle consoled her. 'Instruct your friends not to take mine, and I'll attend tomorrow night's fiasco.' He contemplated throttling her, and dismissed the idea. *How dense can a woman be?* Incensed by her denial of his amorous advances, he gripped both her wrists firmly.

'Ian, you're hurting me. I'll comply with your demands, if you release my wrists.' She twisted free of his hold. 'You're a beast, and a narcissist. Look, you've left huge red welts on both my wrists. Do you enjoy being cruel?' Rubbing them to ease the friction-burns, severe pain brought tears to her eyes.

'Yes. You wouldn't have listened to me otherwise.'

'Why must we keep arguing? I do love you, Ian.' She sidled up to him expecting a cuddle or a kiss. Instead, the fierce glint in his eyes heralded fury.

Peeved, he pushed her away. 'Do you? I wonder! No man should insist on being heard or understood. I hate being taken for granted by anyone, especially you.'

'Darling, don't be angry with me, please? I meant no harm in what I said. Truly, I misunderstood what you were trying to tell me. That's all.'

'If you persist in defying me Gigi, I can force you to respect my decisions. You *are* travelling with me to Zurich, whether you want to…'

'No. I am not. Why must you deny me the pleasure of helping Frau Traugott to prepare platters of delicacies and the filling for savoury tartlets in their kitchen?'

'Right, if that's your attitude…I'll fix you.' Throwing her on the bed, he locked the bedroom door. 'If you don't strip naked, I will do it. How can I make salacious love to you in those clothes? If you object, no one will hear you bellowing. Then I won't feel deprived of your company in Zurich. By morning you might've come to your senses and appreciate how I feel. Remember, I'm no different to other men. We're all susceptible to being hurt by an unkind word, or blatant refusal of a passionate interlude with the woman we confess to love.'

She refused to accept his side of their argument. *Why should I forgive his brutality, after he bruised my arms? It's unthinkable.* In a voice reeking with sarcasm she responded. 'If you apologise about your rude behaviour, I might relent and let you love me, tomorrow night. Now please leave, Ian.'

Her discourteous snarl displeased him. Without yielding, he argued until she finally agreed to spend an hour with him at the ski hut. 'I'll return to my room, if you retract the horrid things you implied. Gigi, don't disappoint me again. Nastiness is an undesirable trait in a young lady. Now kiss me, then I'll leave you to dream.'

Nigh on dinnertime the following night a new Citron cruised to a halt beside the Pension's front garden. Exhausted after a long day of haggling with executive clients, Ian sauntered up to his room. Reorganising indecisive businessmen, who'd thwarted every suggestion he'd proposed had taken a toll on his nerves. With more than hunger on his mind, he allocated enough time to freshen up before visiting his fiancée. His zealousness lessened, even before he stepped across the threshold of her room.

'No, don't touch me,' Gigi wailed as he walked in the door. 'You'll crease my dress or spoil my hair. Have you forgotten we're due downstairs in twenty minutes?'

'No, I have not. Why the attitude? It reeks of sarcasm. If you snarl at me again, you can damn well go down those stairs on your own.' Ian shuffled off to his room. In a huff he thumped the door. It shut with a forceful fling of an elbow.

An inflamed temper smouldered as Ian began grooming his grey hair, tinted with a nut-brown dye. The mirror image reflected a harsh frown embedded on a distorted brow. 'That Welsh bitch is intent on wreaking havoc. She doesn't give a damn about me or my feelings. Tonight when we're alone I'll conspire to bed little Miss Nasty…unless she concocts some idiotic fabrication to deny me that pleasure. The idea of purchasing her an engagement ring in Zurich I abandoned. My deceased wife's emerald ring will suffice. Glancing at his tan-leather valise beside the bed Ian continued to think. *Erika's pearls will enhance that scrawny neck of hers on our wedding day. A simple-minded girl, Gigi will accept my offerings of penitence graciously and without question.*

Precisely at eight, with her face portraying a radiant glow, she tapped on his door. Twice she knocked. No answer. 'Where can he be?' Anxiously she removed one of her elbow-length gloves of cream satin and rapped again. Awaiting an answer Gigi checked the seams of her silver-grey, fifteen denier stockings. They were straight.

The door creaked open. Dreamily, Ian appeared in the doorway, naked with a blue towel barely covering his genitalia. 'Come in, my dear. I've finished shaving. You can sit on the bed while I dress.' He smiled, wiping a wisp of cream from his chin on one corner of the towel.

'Ian, do you realise the time? It's late. I'll be in the recreation room downstairs.'

'If you leave this room, forget tonight. It's your choice.' The click of his thumb and forefinger she took as a threat. Intimidated, she stepped back against the bed. 'How long must I wait for an answer?' This harsh retort startled Gigi. 'I prefer to make love to you here and now, rather than attend their damn soirée.'

Without warning he whipped the bath towel asunder. In a full frontal pose he leered at her. Blushing, she tried not to lower her gaze. He scrutinised her disdainful pout, with disgust and laughed. 'You've seen me naked and cradled my erect penis. Nakedness in its true form is a creation of beauty. Not to be frowned on, Gigi. Look at my physique in this pose of an ancient Grecian God. Originally my maternal grandparents came from Athena. Although my mother was born in Scotland. I learned the art of bodybuilding in a Parisian studio, near the River Seine. Run your hands over the firm flesh of my athletic chest and feel the muscles tense. The pounding of a strong Athenian's heart beats in this athlete's body.'

She shuddered and feeling embarrassed she didn't know where to look. 'Ian, have you been imbibing on some illicit cocktail, or drugs? You're weird! Fantasising! I'm not staying in this room a moment longer. Our guests await the pleasure of my company. If you want to join us downstairs, treat me with respect. Your suit, black bow tie and white ironed shirt are on that bed. Use them.'

She flaunted out of the room with the graciousness of an angelic spirit in flight. Her diplomatic strategy had worked better than she hoped. Carefully chosen words had demoralised a megalomaniac's ego. Crushed under an Athenian's foot! This demigod's self-appraisal lacked the elegancy of his Germanic heritage.

In his distinguished stride, Ian approached the recreation room. His creaseless brow belied the anger building behind steadfast eyes. With a courteous bow and an indignant smile, he nodded to Frau Traugott and then her guests. They chattered amicably until she interrupted their friends. 'Excuse my intrusion, gentlemen. Ian, it might be advisable if you and Gigi take your seats at the top of this table. The musicians are tuning their instruments and their pianist has arrived. You'll find Gigi next door, in the library. I have some idea who the culprit might be…for her tears. Console her. Your charm usually pacifies ladies under duress. She'll hold you in high esteem, if you treat her with respect. Remember, her affection for you is genuine, so approach her tactfully.'

Guilt never weighed heavily on his conscience. In a reticent mood, he listened to her apology. 'Don't let's argue. I didn't mean to hurt your feelings, Ian. Please forgive me, my darling?'

He nodded, but didn't reply.

'You demeaned our love by saying such horrid things this afternoon. They hurt me'. Gigi spoke in a soft voice. 'Why can't you be pleasant for a change? You're always so grouchy and abuse me whenever you *think* I've done something wrong.' Feeling elated after being exonerated, and with her head raised in a haughty manner, she trotted off to mingle with her friends, some of whom had just arrived. A variety of non-alcoholic drinks were offered by the elderly waiter. Gigi chose a lemon squash with a touch of bitters, her favourite aperitif on hot nights.

Selecting an alcoholic drink from the salver, Ian paused until the waiter moved on. A disparaging look framed his downcast eyes as a crowd of well-meaning people hovered around the "honey pot". *Why she bothers with uncouth scum from the streets astounds me.* Ian barged through a group huddled together. He assumed Frau Traugott's friends were discussing his gruff manner. Far from the truth, they were astounded why Gigi wanted to return home to Llangollen, with winter hovering over the Cambrian mountain range.

Proud of his achievement to humiliate her earlier, Ian cruised in a stately manner towards Gigi. In a ploy to ignore his unsympathetic platitudes, she turned to speak with her childhood friend. She greeted Ruffina with an abundance of enthusiastic chatter.

Her rudeness wasn't appreciated. Disgruntled, Ian gourmandised on fresh figs and imported cherries. Hot savouries topped with salmon, oysters and miniature fish cakes and delectable fingers of lobster followed. Small compotes of braised lamb or beef precooked over a spit, topped with parmesan cheese and hot croissants were on the menu.

Gigi gasped when Ian deposited himself in between her and Ruffina at the table, set with Frau Traugott's best china. Silver cutlery and crisp linen napkins took pride of place on the white, lace-edged tablecloth.

Candles flickered and dimmed as a firm hand gripped the nape of her neck. 'You won't elude me again, Gigi. You're mine tonight. If you object or make a sound, I'll twist your hair until you cry with pain. This afternoon you insulted my intelligence by casting dispersions on my lineage. You knew my parent's history and accused me of taking illicit drugs. You also implied I'd been drinking. I'm willing to forget your nastiness, if you promise not to belittle me and my family again.'

She couldn't retaliate or do anything. Within seconds his fingers released their harsh grip on her neck. Falling loose, the upswept ringlets fell down to drape her cool shoulders.

'Come my dear, I'll escort you over to the dance floor.' His gruff manner mellowed a fraction as she accepted his arm. 'That's if we can weave our way through this crowd. What a fracas. It's worse than facing a firing squ...' His soft faux pas was undetectable. *Hell, I can't believe what I almost said. What a stupid mistake!* Concealing his blooper, he counteracted it by saying, 'Darling, let's hope none of the guests have subversive ideas,' Ian sniggered. 'They would be out of character at a soirée of this nature.'

She couldn't see anything amusing in his crass remarks. Angrily she toddled off to the lady's room. *Evidently, he delights in embarrassing me.* Reliving her shame, she recalled having looked at a book lying on his bed. It came as a shock when he snatched the red, leather-bound folder from her hands. Gigi could still hear him saying 'Uh-uh! My private comments and finances are in that ledger. It's for my eyes only.' Redeeming it, he stuffed it in his briefcase. On recall of that incident she frowned when he said, 'This valise of mine is another thing you do not touch. Can't have all my business affairs splattered all over town. Ask me in future…before you go snooping'.

A delicately scented handkerchief now caressed her wet eyes. Sniffing she thought, *I tried not to blush when he forced me to squeeze between his naked body and the door. I think I was still shaking when I entered my room.*

Intrigued by her bemused pout, on her return from the toilet, he nudged her arm. 'I just spoke to you Gigi and you ignored me. Why! Were you thinking of our last quarrel? I told you to retouch eye-eyeliner in there. You snubbed me and sarcastically inferred that my family's history was fictitious. I will forgive you…if you promise not to misbehave again and you treat me with the respect I deserve.'

Although she protested, he insisted they should dance. In a regal pose she glided to the lilting rhythm of *Paris in Springtime* as the silk-taffeta skirt of her ebony gown rustled. Ian stepped behind her and whispered, 'I'm warning you not to discuss last night's argument in front of your friends. What I say to you in private…is none of their business. You know my temper is aroused easily.'

On reaching their seats she paused. Perplexed, Ian lowered his eyes and scowled, first at her, then the crowd. 'This was supposed to be an informal gathering? Just a few friends, you said.' In a rancorous tone, his voice alternated between its lowest ebb and a high-pitched drawl. 'Why didn't you have the decency to warn me?' It equalled the tone of her gasp.

'Darling, please don't raise your voice, or embarrass me in front of these people. If you spoil my evening, I'll leave you for…ever. And I mean it, Ian.'

'No, you damn well won't leave me, or this shemozzle. I intend to forget what you just said.' The snarl was indicative of his mood. 'I will *not* allow you to cause a scene or make a show of *me*. Smile and get rid of the sullen pout.' The arm he held was pinched.

Gigi reciprocated with a growl and edged away from Ian. Concealing her deflated ego and hurt feelings, her forced smile imaged a gasping ape.

'What a hell of a crowd. I'll speak to you later about this. It's unnerving.'

Frau Traugott, having noticed the couple in silent conflict, moved in to request, 'Will our guests of honour be seated in their original seats.' In a courteous gesture she assisted Gigi to hers. 'Darling, has Ian told you how beautiful you look this evening?'

A forced grin masked her anguish. 'Yes, twice on the dance floor, Frau Traugott.' Intentionally directed at him, this lie had escaped though pursed lips. Humiliated by his nasty inferences how could she disclose the truth behind their discord?

By hugging the lapin stole tightly, it concealed her bruised, shivering arms. She wasn't cold. The euphoria of descending the stairs in a graceful manner had long since dissipated. Now she felt drained, depleted of energy. Only a dismal remembrance of her grand entrance remained. Finding it difficult to ignore his draconian attitude she presented a brave face. With a pounding heart, she feared this overassertive and possessive man by her side might pass more undeserving, critical comments. With those close by observing her, Gigi spoke in a ventriloquist's pose, as one corner of her mouth drooped slightly. 'One more snide remark, Ian and I'll withdraw my promise of marrying you. If you disapprove of my friends leave now. The choice is yours to make. Not mine to decide.'

'Yes, I do object Gigi. That couple standing by the piano. They're garish and loud-mouthed fools. Their voices should be lowered to a nominal tone.'

'Now, it's my turn to ridicule you. They're strangers who've dropped in to ask Frau Traugott if there's a spare room here overnight. You should be sure of your facts, before you condemn innocent people. I know you didn't mean to be rude. Please try to enjoy this evening by not making more insulting comments about strangers and my friends.'

'You're wrong, my dear. I am not discourteous.' His muffled rebuke came across as insincere. 'You caught me unawares. Remember, you're used to crowds. A conservative man, I'm not.' Ian thumped her arm.

'Gigi, you'll not embarrass me again. If you do, we'll go upstairs. That infantile pout of yours is soul-destroying. I disapprove of it. Look in your compact mirror and you'll see what I mean.'

She didn't answer. *Perhaps I may've been a bit hasty and jumped to the wrong conclusion? Why is he being so damn difficult? He doesn't understand how a woman feels when she's under stress.* Passive thinking conflicted with her benevolent ideas.

Feeling like a stranger among a lot of tittering females, who congregated in groups with their husbands or partners, Ian scoffed.

'I am seldom irritable. All this nonsensical hoo-ha is unbelievable. I'll let you know when I tire of this group of ignoramuses.'

He squeezed her hand. Repulsed by his crude comment she lifted the offending fingers and let them drop in a bowl of warm soup. Offended by his obscene remarks she stood and walked off to confide in her closest friend. Ruffina sensed her anger, it showed in Gigi's eyes. Nothing more was spoken of Ian's arrogance, and belligerent behaviour or his uncouth display of rudeness.

The evening proved a success and was enjoyable. With an afterthought, Ian pretended to see the folly of his nasty appraisal of her friends and their guests. Twice he passed complements on the graceful way Gigi moved in her crystal-beaded gown of burgundy silk, shot with a silver glow as she danced.

Ian overheard someone comment that he indeed was a fortunate man. Big-noting himself, he made it obvious by declaring to his table-neighbour, 'Gigi consenting to be my wife, her future will be assured. With my high aspirations for our lives together, she will never need to stray into that Frenchman's arms.' Cautiously he tried not to overdo the accolades. However, his sly dig about Pierre Bouvier hadn't gone unnoticed by Frau Traugott. Excited about the promise of an engagement ring and twittering to Ruffina and their friends, Gigi had missed Ian's snide dig at her French tutor.

On conclusion of supper Ruffina nodded to the ladies-room. Something had cropped up and she needed to discuss the problem with Gigi.

'What is it, Ruffina? Not your husband again surely...'

'No thankfully, he couldn't attend tonight's soirée. Maurice finally knows I won't tolerate his abuse any longer. This is not about him. Don't be annoyed Gigi, but I've done an awful thing. And I may lose your friendship and respect.'

'Words only hurt when they're said in anger. I won't be angry with you, because I value our lifelong friendship. What is worrying you, Ruffy?'

'Actually, it's your fiancé I owe an apology to Gigi. I took two group photos…'

Gigi's laughter created a stir in the busy toilets. 'Yes, I know. I'm in two of them.' It dawned on her about Ian's insistence of no photographs. 'You took a photo of Ian,' she whispered excitedly. 'How wonderful. I won't tell him, Ruffy. It'll be our secret. Don't destroy the snaps or negatives when that roll's developed. As yet I don't where…or what our address will be in London, or in England.'

'Not to worry, darling.' Ruffina unclasped her beaded evening purse. It contained a book and fountain pen. 'Manny's address in Llangollen is written here, in my diary.' Her camera found its original niche, in the pocket of her evening clutch.

Gigi winked her long mascara-tinged lashes in a return gesture. 'Don't forget to register the films, Ruffina. I'd die if those photos or their negatives went astray. Our mailman is usually reliable. Last month the local kids stole Manny's letters just to see if they contained money.' They both knew her grandmother would forward the snaps to Gigi in London, or keep them until she returned to Llangollen.

'I'll be home for six weeks, maybe longer. It depends on Ian's workload in London. Ruffina, address your letter to me there. Remember, it's our secret, so don't tell your husband. You know how men congregate to discuss their wives' inadequacies. Ian will kill me if he finds out you've taken a snap of him. Let's enjoy a drink with our friends, or the night will be over.'

Exiting the toilets Ruffina sniggered, tilting her head she pointed. Standing behind her was Ian, who couldn't see the shocked look on Gigi's face.

As she turned he grabbed her hand. 'I've already said goodnight to your guests and the freeloaders. Frau Traugott has explained to your friends that you're not well and why we need some privacy.'

Without warning a sudden urge to leave caused Gigi to clasp her forehead as she swayed. 'I feel as if I'm going to faint…' As she tilted forward Ian caught her. Shocked, he saved her from collapsing on the floor.

Frau Traugott pointed to the stairs. 'Carry her straight to her room Ian. I'll be there shortly.' Instead of a creating another unpleasant scene, he acknowledged her advice with a grunt and a sly nod, which meant *move and where else would I put her?*

Ruffina collected Gigi's purse and followed them upstairs. 'I'll help her to undress. Frau Traugott, she looked very pale before she fainted. My husband won't be here for ten minutes. She'll be safe in my hands. Will you call me when he does arrive, please?'

Frau Traugott nodded as Ian carried Gigi up the stairs. Ruffina rushed on ahead to open her door. Folding back the covers she moved back to let Ian put her on the bed. 'I'll call if I need you, sir. Gigi needs peace and quietness to recover. She's gone through hell lately. Now, will you please leave?'

Begrudgingly he strode along the corridor grumbling about her rude request, after indignantly being ordered from the room.

'We deserve time alone without that bitch interfering. She knows why Gigi's ignored me all night. Surely, she's not still annoyed by me chiding her. Well, she needn't think I'll be her slave. I wanted to discuss our future together. We can't push our problems into the backblocks of our minds forever. I *am* worried over her health. No, concerned if she's changed her mind about our marriage.' Altering his stance Ian thought, *I expect news of a good job, teaching youngsters in the Cotswells. It'll enhance my prospects of obtaining a tutoring position in a London University. We'll need money to live on, and I won't lose this chance to be a multilingual lecturer. English and German are just two subjects that I excelled in at Edinburgh. Gigi doesn't know I obtained honours in three other subjects and that I speak French fluently, plus a smattering of Russian.*

An hour after Ruffina returned downstairs, Ian approached room nine, two away from his. He gave a gentle rap on the door.

'Come in Ian. I'm tired. I'll be fine in the morning, if I'm not pestered and have a reasonable night's sleep.'

'What did the doctor say caused your collapse? I didn't mean to offend you, but you refused to listen to me about our marriage being held in London.'

'We can't leave for England, not now. I must go home. Manny will be disappointed if I marry you, without seeing her first. An elderly lady, she'll rely on my help until I can arrange a live-in nurse. I'll make the arrangements to leave here, then let you know. I have a little wrinkle up my sleeve. Don't ask why, or how. Be patient and wait, Ian. I'll give you an answer tomorrow, or after I hear from my business friend in London.'

'No, I've made up my mind,' Ian adamantly insisted and rambled on regardless with his ideas for their future in England. 'Did you tell that friend of my plans once we land at Gatwick? I can't afford anyone to dissuade you from coming with me as far as London. Then I'll book and pay for your airline ticket to Manchester. You'll have to make alternate arrangements from there to Llangollen.'

Ian's deliberate ploy to deceive Frau Traugott had worked. She thought they would be spending the day together hiking. She knew nothing of his intended journey to Zurich. This lie he'd instigated to give him time alone with Gigi. His insatiable desire to love her couldn't be restrained. Ian knew they wouldn't be disturbed at the ski hut.

Now with his lustful desires reaching fever-pitch, he donned a shirt and casual slacks for tonight's sexual liaison in her room, without being disturbed by anyone.

8 pm: On conclusion of consensual petting, they consummated their love. Still in a close embrace they quietly discussed the evening until he presented her with a double strand of luxurious pearls. Naked and unashamed, Gigi paraded in front of her mirror to appraise their luster. Adjusting the strands over her cleavage, she caught Ian peeping at her naked buttocks. Embarrassed, she blushed.

In her mirrored reflection his eyes focused on everything, but the pearls. Desirous of loving her again, he coaxed her back to bed.

'My dear, remove the pearls. The waxed string may break if you pull on them. His voice was harsh yet it held a gentle connotation, as he thought: *They cost me a fortune in Hamburg during the war. I can't afford to buy another set, if she snaps the cord.*

9 20 am the following day. They were ready to leave the Edelweiss Pension when Frau Traugott thwarted their plans. As Ian turned his key in the ignition slot she presented Gigi with a basket full of wine and delicacies.

'My dear, I thought you might enjoy these goodies in Zurich. I'll understand if you aren't home by dusk.' As the frau smiled Ian detected a wicked glint in her eyes. 'If you're not here by six, I'll keep your dinners warm. Enjoy your day together. Drive carefully, Ian. A man on the radio just reported a landslide near Zurich.'

Proud of her deceitful ploy, she smirked like a mischievous child. On their short drive to the ski hut, Gigi wondered if they had fooled their hostess. Excitedly her feet tapped to a mystical tune on the carpet. Parking his car in protective shade, Ian unloaded everything as she waltzed to the musical tune her mind had conjured and she wandered on ahead. He warned her to be careful not to fall in a puddle or slip on the wet grass.

Yesterday's rain had dispersed and the sharp morning wind had dropped to a whispering breeze. This unseasonable blast gave the impression that the bleak winter may linger right through until early spring.

'*What a sacrilege to remain indoors on such a glorious morning,*' she trilled. '*Dilly-dallying around the old ski hut will be boring. Love can wait.*' Leaving Ian to his own devices, she toddled down to the woods.

However, his idea of spending the day together didn't include wandering under wet trees or through dense foliage. Hurrying, he caught up with Gigi by the woodpile. Without a word he tossed her over his shoulder. In an uncivilised manner Ian carried his wriggling bundle inside the hut. A foot slammed the door.

Caught unawares she was furious. Sanity ruled. She decided not to make a fuss of their unspoken conflict. Although she did welcome the idea of being together, instead of a furious tongue lashing...which she expected.

Settled on the woollen rug in front of a warm fire, he tenderly endowed her with gentle foreplay. Naked, he held her close. An hour later after another passionate and romantic interlude they dressed. Gigi turned away from his gaze to put on her cream satin scanties and petticoat. Donning her autumn-toned chiffon frock she turned to confront Ian with a seductive smile. Frau Traugott's delicious sandwiches and lemon tarts went down well with apple cider. Munching on apples, they strolled along the river bank.

Deliriously happy, she enjoyed the tranquil hours in privacy to talk with Ian about their future together which now seemed a little uncertain.

Intuitively, she began to doubt his sincerity. Gigi sat on a log, free of damp lichen or weird little creatures, to admire the brilliant sunset. Her eyes focused on the ball of flamingo-fire as it dipped below the horizon as Ian held her hand. Turning the palm upwards, in a moment of tenderness he spoke in an endearing, though calm tone. 'Close your eyes and don't peep until I tell you. Gigi, I saved this precious gift until now. It represents our future life, bonded in love.'

She imagined herself dressed in a Cinderella gown with Ian as her Prince Charming.

'Oh darling,' she gasped on opening her eyes. 'This emerald ring is beautiful! It must have cost you a fortune.' Unobtrusively thinking she wondered why he didn't answer. *Perhaps he's embarrassed by my intrusive comment by the cost of this ring.*

Gigi proudly positioned her hand so the gemstones would catch the cosmic rays of a dying sun. Excited, her eyes followed an array of hews reflecting off the gems onto her sheer, full skirt. With every twitch of her finger, flashes of green to fiery red reflected on the still water, it reminded her of fairies dancing on a lake in the coral flush of twilight.

'They're real emeralds and the ring damn near broke my pocket. It's a perfect fit, my dear.' Ian embraced her as they ambled back to the hut. 'You can look at the ring later. Gigi, doesn't a man deserve something in return…for the gift?'

She appraised his downturned look with adoration. 'Do you mean a kiss Ian?' With a charismatic smile, she paused with pursed lips to admire the glorious sunset.

Annoyed he gasped. 'Definitely not. A paltry kiss for an expensive gift I consider an insult. Passionate love is what I have in mind, before the sun fades,' he confirmed as hand-in-hand they re-entered the ski hut.

Before she could respond, his lustful desires consumed him and in a passionate embrace he made love to her on a rug in front of a fire's dying embers. Closely entwined and linked as one, they dozed in ecstasy.

The cry of a night owl disturbed their peace. While dressing they discussed various topics of mutual interest as his mind lingered on her sudden departure days before from the Edelweiss Pension.

'I suppose you wondered why I left there so soon after my accident.'

'Yes, I have numerous times. Gigi, the note you left with Frau Traugott I considered offensive. I didn't appreciate being accused of forcing my attentions on you. Darling, you must have sensed how I loved you, even then.'

'You're not going to agree with what I'm going to say now either. I'll tell you the truth why I left. It wasn't your fault, Ian. I received a cablegram from France. Marion, my friend Pierre's wife has been diagnosed with cancer.' A hand gesture warned him not to interrupt. 'I was uncertain whether I should go to Paris, or stay here with you.'

'So you do have feelings for *that* Frenchman. What did you intend doing? Wait in the wings for him to make a move, once his wife died,' he snarled, sitting naked on the bed he watched Gigi clip the fifteen denier stockings to her suspenders.

His uncouth comment struck as if a knife had pierced her heart.

'Ian, how can you be so insensitive. Of course not,' she snapped back. 'I love Pierre and his family. They're like my own kin. They've been very good to me down the years. Pierre's a father figure to me.' Moisture clouded her eyes and she brushed a tear from her cheek. 'You must believe me darling, it's you I love. Please don't be nasty to me, or Pierre. He deserves to be treated with respect.' She paused to gather the random thoughts rampaging in all directions through her mind. Fearing he might retaliate in some way, she stepped well out of his reach and sniffed.

For a change Ian kept quiet, although he was tempted to intrude. Hesitant, he let her wipe the tears from her dark eyelashes. *There's nothing more depressing than a maudlin or weeping woman.*

'Why don't you think of my feelings, instead of your own selfish ones for a change. Pierre was my ballet teacher. Now he's my business partner.'

Pursing his lips Ian didn't reply. The idea of another man accepting her love, burnt deep in a distorted mind.

'There's one thing we need to settle while we're on the subject of partners. I disapprove of you working in *his* ballet company...or any other venue, once we're married.'

His motive was to segregate Gigi from opportune meetings with the Frenchman and their friends. There would be less chance of their marriage failing. Whenever Pierre Bouvier's name cropped up, it felt like a bomb exploding in his brain. Ian twisted his face in such a distorted manner it reminded her of a monkey she'd seen in the Parisian zoo. Rekindling the fire he frowned. *Once we are married, there'll be no room in our lives for that French bastard or his family.*

A lingering silence occurred until Gigi intruded on his solitude. Clasping both his hands she gazed up into his cold, ice-blue eyes.

'Darling, we must rearrange the company's finances. Pierre intends to let our understudies take over each performance. They're both talented dancers and quite capable of handling all press-reviews.'

A deep furrow divided his eyebrows. Ready to challenge this ridiculous notion, he froze and momentarily refrained from speaking.

'Ian, will you listen to me? I have arranged to meet Pierre in his solicitor's office in London. We need to sign the final legal paperwork for us to receive the royalties. At this stage, how can I disappoint him?'

Ian stroked her fingers and plied their tips with a kiss.

'No Gigi, you listen to me. I've come to an important decision. I propose to let you use my postal box in London. Once we're settled, I'll arrange for my solicitor to draw up the papers pertaining to your business deals. I can't be fairer than that, my dear.'

'What use will your postal box be to me, when I'm in Wales? Don't let's argue Ian. Things will work out in time, I guess.'

'No, I want to hear nothing more on either this subject...or him. Would you prefer me to contact Pierre personally? Unless you get rid of these silly notions, I'll make it my business to phone Monsieur Bouvier while you're away and state my displeasure. Then we'll sort out this mess once and for all.'

Hurt by Ian's jealousy and mistrust she was incapable of expressing her frustrations. *I'm not a fool. It's blatantly obvious what he means. The horrible situation will escalate if his negative attitude worsens, especially if this controversy continues.* On their return to the Pension she remained pensive. Gigi resented his persistence to revive their previous confrontation.

'Six weeks without you will seem an eternity to me.' This declaration possessed a contemptuous overtone. 'It'll be sheer hell here without you Gigi, while you're enjoying yourself up in Llangollen. God knows where I'll be? Lost and feeling lonely, I suppose.'

How could she respond? Infuriated, she stepped from the vehicle as it cruised to a stop outside the Edelweiss Pension.

An hour later the same blank wall of silence persisted between this feuding couple. Gigi was the first to back down.

'Let's be friends, Ian. I'm sorry if you mistook what I meant. You were nasty and hurtful. Please forgive me. My lack of knowledge in your business affairs was because I'm a novice to the field of manufacturing war machines…'

'It's not before time that you acknowledged being foolish. I resented what you implied. Never mind. It belongs to the past. That argument won't be repeated. I have forgiven your response to my civil demand. It's time we went indoors. Frau Traugott must be curious why we've arrived home late after a day trip to…Zurich.'

Wera Traugott anxiously awaited their arrival and by nine she began to panic. Hearing the car, she rushed to greet her guests. Flagging one hand she gasped, 'Thank goodness you're home. This cablegram came an hour ago. Gigi, you should read the contents now. Come through to my office.' The frau hurried on ahead to her office and gestured for Gigi to sit opposite her at the desk.

'Ian, I sent for a bottle of our cellar's best champagne.' After an earlier incident, her frown warned him not to imbibe heavily. 'You'll enjoy a sip in your room tonight.'

'Wera, is the cable from Pierre?' Gigi pleaded, stammering nervously as she fiddled with the envelope in her sweaty palm. 'I…I dread the idea of terrible news.'

'No, and it's not from your Grandmamma either. I must admit we will be sorry to see you leave here. Read it Gigi, before you misjudge the meaning.'

She glanced at the typed text unsure what to think. A fervent scream escaped from her pouting lips and the gram jettisoned over their heads.

'We have it, darling. We leave in two days for England. Yippee!' Her raised fist deflected a warm breeze whistling in office window. 'Ian, see what knowing an important diplomat and a British charge-de-affair does? Positive results every time.'

Astounded, he didn't approve of her sarcasm, nor her over-enthusiasm. *This news is a godsend in disguise. It passes my expectations. My contrived plans have proved their worth.* Ian's resilience mimicked his enthusiastic elation.

The following day flowed without dissention. By nightfall Gigi tried ardently to dissuade Ian from holding a grudge against her and Pierre's long association in their field of arts and dancing. Stubbornly he refused to listen.

4

On a magical May morning, Herr Traugott drove their guests to Zurich Airport. The elderly man would miss Gigi's exuberance and her pleasant smile, although he welcomed her suggestion to keep the Citron. A responsible vehicle, it far outstripped his dilapidated wreck. Before they left, Frau Traugott tried to persuade her departing guests to return to the Edelweiss Pension the following spring. She even offered to cater for their wedding. And suggested they could stay *a gratis* for their honeymoon.

Gigi seemed pleased with these arrangements. However, Ian wouldn't have a bar of her idea. Firmly he stated *NO*, adamant that their wedding would be in London, where his business would be easier to conduct. The decision made, he now stood by it. Naturally she was disappointed. Finally, in her room Gigi relented. *It doesn't matter where our nuptial knot is tied. In time Ian may relent. He is a stubborn man at times.*

On entering Zurich's International Terminal, he hurried to collect their prearranged flight tickets. A hand gesture hailed an athletic young porter to carry their heavy luggage over to the departure desk. Ian gave the lad a small tip for his trouble.

Left to fend for herself, Gigi battled with an awkward suitcase, her hand luggage, woollen topcoat and purse. Depositing the heavier items by the airport's main entrance she sat on her case until Ian returned.

Mr Homer Ellis, London's embassy Chargé d'affaires' secretary, in the process of searching every annex caught her looking lost and miserable. In a dignified manner, he approached his lifelong friend.

'Gigi, it's terrific to see you again this soon.' They embraced. 'Allow me.' Homer collected her heavy suitcase. 'I've been hunting everywhere for you. I thought I might've missed you in this crowded complex. Darling, you won't be permitted to fly without completing this paperwork. My fault, I confess,' he said handing her a form. Listening, she glossed over the brochure. 'Gigi, I forgot to ask you to sign that document the other day, after I returned your passports.'

Her palm graciously awaited his pen.

'Homey, you mentioned at the time there was something you needed to witness our signatures on. That happened after we signed all those declaration forms.'

'My dear, I specially came here to give you this document,' Ellis stated. 'Sign it now, and I'll leave you and your fiancé to commiserate.'

'Commiserate! Why?'

'The media contingency's special flight to England, your flight, has been delayed for several hours…as it stands.' They sat down together on a nearby seat while she filled in the page as Homer observed a huge backlog of luggage rumbling along the carousal. 'Gigi, I tried to arrange for you to travel on a commercial flight. No go, I'm afraid! Everything's booked solid with people attempting to get home, away from Zurich.' Homer's eyes embraced the hordes of dismal faces hovering in front of schedule boards. 'That rotten war has torn their sad lives to shreds. Here in Switzerland, we must continue to cater for the desolate refugees and their needs.'

'The war's been over for months. I never imagined the airport would be so busy. We were lucky. Pierre and I arrived here in the dawn hours, as you know.'

'How is Pierre? I was hoping to meet him again. Where is he now, Gigi? Has he returned to Paris to be with his wife?'

'Yes. He only stayed at the Edelweiss Pension for two days. Marion's ill and you know Pierre. He needs to be with her. Homer, I do and always have admired his loyalty.'

'You'll never get over his charismatic charm…if you don't really try. A man of his stature never changes. Why aren't you honest with him? Tell Pierre how you feel.'

'No, I have to forget how I've felt in the past, and try to put him out of my life. I do love Ian. In a different way, though. Homer, I desperately want us to be happy. And I'm sure Ian feels the same way.' The final document was signed as she spoke. 'Does he have to sign this one?'

'Yes, Mr Ross does. Point out the direction he walked and I'll take it over to him. That will save all of us time.'

'Homey,' her fingers gripped the tweed-jacket. Please don't say a thing to Ian about Pierre, or my feelings for him. As yet he hasn't met him and he's really antagonistic towards him. Ian flares into a violent temper every time I mention Pierre's name.'

'Gigi honey, are you sure you're doing the right thing marrying Ian? I'd hate you to waste your life, especially if he has a jealous trait. Please give it some thought…a lot of thought. We all love you. I fear you'll be making a mess your life for the sake of waiting a month, or two.' In a gesture of kindness, he gently touched her hands. 'Are you going straight through to Llangollen, or stopping in London first?'

'Bjorn and Kirsten have offered to put us up for two nights. Ian told me on the way here his mother is ailing. From what I can gather, he must've been born when she was in her mid-forties.' Anguishing over what Homer had said Gigi scanned the sea of faces congregating around them. 'Ian's flying to Scotland on Thursday at noon. He told me in the car, a friend in London made those arrangements yesterday.'

'Give the Svenssons our love. My wife wrote to Kirsten last week, so she informed me this morning at breakfast.' As he spoke Homer, a tall man, peered over the great expanse of heads moving in all directions within the airport complex.

That ignorant bastard has held me detained long enough. There's a lot of official paperwork awaiting at the office. Homer Ellis thought it necessary to discuss several important issues with Gigi. There wasn't time and he looked puzzled over her insistence to marry a man she barely knew. Instinctively, he sensed she would be in for a hell of a life with Ian Ross, whom he considered an arrogant man, and disliked immensely.

'Tell me one thing, sweetie? Why did your fiancé wait this long to tell you about his ill mother? I would've told you instantly.'

'Homer, you're not Ian. My man has his own way of dealing with urgent problems and I daren't interfere.'

'Have you met any of his family? Gigi, I asked that because you mean a lot to me. I'd hate to stand idly by and know you're making a huge mistake. In my opinion you may later regret rushing into a marriage with an argumentative stranger.'

'My love for Ian is genuine. Ha, I can't talk, not with my temper…'

'You and a temper?' he interrupted her and laughed. 'Come now, you forget this is Homer Ellis you're talking to Gigi. Don't forget, I've known you all your adult life. What am I going to do with you, my girl?' The attaché's secretary wagged a persistent finger at her. 'Oh, I almost forgot. Sharon sends you her love. And regard from my boss. Here…' Homer took a card from his grey vest pocket, '…keep this out of Ian's sight. Make sure you can find it…and he can't. Should you need Henry, or myself you can reach me twenty-four hours a day on that number. Then I'll relay your message to the Chargé d'affaires.' Accepting the card, she vaguely gazed at it. 'Reverse the charges. Quick, put the card in your purse. Your fiancé's walking this way.'

Gigi tucked the card in her handbag then turned to face Ian. 'Darling, Mr Ellis has just arrived from his office. Ian, you must sign this paper in front of him. I've already signed there. That penciled cross is your spot.'

After scrutinising the item in his hand, Ian glared at Homer. 'I was under the impression that we signed this document before.' His dislike for the man whom he was addressing accentuated the cruelness in his eyes and also in his abrupt manner.

Mr Ellis moved in between the couple and politely eased Gigi back. 'Excuse me Mr Ross…sign there.' A manicured finger indicated the spot. Handing him the pen Homer smiled. 'Then I'll leave as I'm already late.' He watched Ian inscribe his Scottish non-de plume. 'That's fine, thank you sir. As I've already explained to your fiancée, the flight you'll be travelling on to Britain has been delayed for several hours.'

Receiving the signed document, Homer acknowledged Ian's grunt. 'I'm delighted to see you again, Gigi. My wife will write once you've settled. Let her know your new address please.' He held out his hand to Ian, who rudely ignored both it and him.

'Congratulations, you've won the heart of a beautiful lady. Look after your wife Mr Ross. She's precious to us. My wife and I consider her a special person in our lives. I trust you will be happy in your new life together.' Homer remarked on the quiet, as he strode past Ian. 'Don't let me hear of you laying a hand on, or hurting Gigi. If you do sir, believe me…you will be sorry. I mistrust all fools.'

'Sir, I realise how fortunate I am.' His sharp retort retained a low timbre. 'Come darling, I found a locker to house our cases. Goodbye

Ellisss,' he hissed under his breath as they walked away, 'and good riddance.' Carrying their cumbersome luggage, he gave Gigi's nearside arm a hard nudge to make her hurry.

Looking back, she waved to Homer Ellis, with the coat over her free arm. Parting in such an abrupt manner was a sad reflection of their long friendship. She wanted to say goodbye with a hug. She knew Ian would misconstrue this display of genuine affection for her friend.

Observing the couple leave, Homer Ellis focused on Gigi through mist-laden eyes. Departing the terminal, he waited under an awning until his chauffeur-driven limousine eased kerbside. Cursing, he rechecked his wristwatch. 'Damn! Now I will be late for my appointment with the British Consul.'

Homer's prayer then came in the form of an exasperated sigh. 'Please God, keep her safe. Gigi's in for a rough time with that bastard. I'll kill him if he raises his hand to her,' this hushed murmur exited through taut lips. 'Then he'll answer to my boss and Pierre. They can stand in line, after Bjorn and I have finished with that pompous prick.' About to enter the vehicle Homer glanced over one shoulder. *Your sort Mr Ross, are the scourge of the earth. You're not good enough for Gigi by far. She deserves a man like Pierre…a gentleman who respects and loves her.*

In the departure lounge the future Mrs Ross remained silent. She refused to speak to Ian who had rudely insulted her dearest friend here at Zurich Airport. Homer Ellis and his boss had put themselves out to assist them in every way possible. Now Ian had repaid their hard work with pomposity and arrogance.

The last nasty rebuke displayed his real temperament. *Am I making a mistake? Will Homer forgive me for allowing Ian to act in such a despicable manner? Every word of what I'm surmising, he'll be thinking worse.* These constant ramblings tore through the delicate fabric of a tormented mind. Clutching her aching temple, she looked at the crumpled item in her left hand. Frustrated and in anguish, the temptation to destroy Homer's address-card grew exceedingly stronger. Something made Gigi stop and think. She reefed a creased handkerchief out of her clutch bag to wipe the black-streaks of mascara from her flushed cheeks.

Ian's rude snarl at me for offering to mind his valise, while he bought a paper to read on the plane was unforgivable. His brusque response she considered an insult - "This briefcase never leaves my side." *He warned me once before not to touch the dash thing. If he keeps on being nasty, I'll ignore him.* Gigi's vow of silence ceased, when a woman cut across her path and almost knocked her over.

In a flurry of confused bewilderment, she entered the powder room. In solitude her emotions could be expelled in privacy. 'Perhaps I'm overtired. It's wrong of me to blame him for my shortcomings. I know he didn't mean to be hurtful. He does love me,' her soft monotones came in spasms. 'Stop this rot now, you silly woman. You can't let him see that you've been stupid enough to blubber.'

Pounding the washbasin with both fists, her pitiful sighs lessened as a young woman ushered her child in the toilet bay. Retouching her make-up, Gigi powdered the dark circles under her eyes and swept a few stray strands from a hot brow. After relieving an overburdened bladder, she walked from the toilets to confront her fiancé again.

Busy with his nose entrenched in a newspaper, Ian didn't raise an eyebrow. Finally, he looked up, said nothing and diverted his gaze back to study the morning news.

If he ignores me, I'll pretend to rebutton my gloves. Sniffing, she tried to look casual and leisurely sauntered to the flight clerk. 'Excuse me please Miss, when is this flight due to leave?' Gigi placed her ticket on the desk.

'Madam, please wait here. I'll just be a moment.' The woman walked away, spoke to a male flight attendant then returned. 'They're ready to call your flight now. Collect your luggage and one of these gentlemen will show you where to go. You realise you will be travelling with male service personnel?'

'Yes, my fiancé is with me. We both have these special tickets. Our heavy luggage has already been booked through here. Madame, I need to show the airhostess this letter on boarding the plane.'

The desk clerk looked at Gigi's blue hand-scripted envelope, nodded and passed it back. 'Please Madam, come back here in a minute and you'll be attended to instantly.'

Waving the letter in her hand, Gigi rushed over to Ian. 'Darling, we're to fetch all our hand luggage now. A stewardess is going to let us board in five minutes.'

Indicative of his sour mood he grunted. 'Don't move and keep my valise between your knees.' Displaying an arrogant gait, he strode to the lockers, several paces from where they were sitting. A bitter twist framed his mouth as he half-turned. 'I'll not be long here, Gigi. Don't you speak to anyone or move.'

'Why is he angry? I've done everything he's asked.' Anguished, her pleated brow matched his scowl. 'No, I must be imagining his nastiness. I'm sure he loves me.' To an eavesdropper, her droll-toned growl sounded as though she was trying to convince herself to believe in positive thinking.

Paged to attend the desk, she hesitated until Ian returned with their hand luggage. They proceeded to the designated departure gate. Boarding passes checked, an attendant directed them to the air-bridge.

Afraid of being demoralised by glares from senior officers, and their underlings in uniform, Gigi shuddered on stepping aboard the DC3.

'We're the only civilians, apart from the contingent of Army media on this flight from Zurich. The desk-attendant warned me we would be travelling with British servicemen on leave. Ian, I'm the only female amongst all these men. How embarrassing.'

'You'll be fine with me beside you, Gigi. Their honour is important. They won't take liberties to gawk at you.' Ian looked at her deep cleavage. 'Not that you wouldn't attract any man's attention…with such a revealing neckline.'

Sneaking another peep at her heaving breasts quelled his lustful desire to fondle them. He took a handkerchief from his grey vest pocket and passed it to her under his overcoat. 'Tuck it down your blouse, and ease a corner to cover your cleavage.' A trained tactician he exercised his options to cope with all difficult situations. 'Gigi, turn towards me. I'll distract their attention by dropping my coat. That'll do the trick.'

She nodded, though didn't speak. *My cheeks feel hot and flushed.* Grimacing, she tucked the lawn fabric between the swell of her breasts.

'Your cheeks are flushed. I'll adjust the handkerchief.' He glanced down at the delightful sight meant for his eyes only. The hanky was obvious, but served *his* purpose.

Within minutes their aircraft ascended high above the city of Zurich and then continued on its flight path to London. To retain their privacy, Ian held her hand under his serge coat, now cloaking her knees. Similar to their seats, the opposite row was anchored to the fuselage and faced inward, with a wide aisle between them.

He squeezed her hand and called her an endearing term, it sounded like heartsie. His face blushed red as hers had done a minute ago.

Although he didn't repeat the word, her mind conjured a million names she wanted to call Ian, due to his crassness and offensive behaviour to Homer Ellis at the airport. None of which Homer would approve. Words a lady would never use or repeat in company.

On their way through Customs at Heathrow, she heard her name being paged over the intercom. Gigi appeared shocked as a couple approached them. Flabbergasted, she graciously greeted her Australian friends. Her delight was short-lived. Ian moved back, scowled and with both arms folded, waited to be introduced.

'Gigi darling, you look marvelous. Unlike me, you never show your age. Suppose it's because of all the bull...'

'Don't you dare say it, Bjorn Svensson. We *are* in a public area,' his wife scolded. 'Gigi is a lady and she's not used to your crude Aussie slang.'

'Darls, I only intended to say bull-dust. Not what you thought...'

Gigi interjected. 'How do you like being elevated to flight captain? Bjorn, I hear you'll soon be the general manager of British Airways.'

Kirsten tugged her husband's shirtsleeve, then peered over her shoulder at the tall distinguished gentleman standing directly behind them.

'Hello! I'm Kirsten. Well...Kirsty to my friends. You must be Gigi's fiancé,' she smiled holding out one hand in a gesture of welcome.

Ian smirked. Begrudgingly, he barely touched her fingers and nodded.

'Sorry old sport to leave you standing there like a shag on a rock. I'm Svensson. I gather you're Gigi's other half.' A quick retraction of "*better* half" prevented a row, as a courteous hand protruded towards Ian, who ignored it and Bjorn.

'Svensson, I am *her* fiancé. None the better for your asking. You can assist us with our luggage, *old* man.' Ian placed the emphasis on old. *Who is this smart-arsed bastard who had the audacity to address me in such a tawdry manner?* These thoughts were unexpressed and accomplished by a critical snigger.

'Why, yes of course Ian. Sorry, my mistake. You girls can catch up on the gossip later. Kirsty, help Gigi with her things, while I bring our car around to the main entrance.'

Ian Ross stated, in his brusque and intrusive manner, 'I'll accompany you Bjorggg. The women can take their time meandering to the front of this monstrous establishment. If that's where *you* propose they should wait?'

Gigi knew this mispronunciation of Ian's was deliberate. Quickly she stepped in to say, 'Bjor*nnn*, I can carry my own hand luggage. Thank you Bjor*nnn*.' She then turned to Kirsten. 'I'll follow you in a minute. Ian's forgotten my hatbox.'

'No arguments from either of us, sweetie.' Bjorn smirked over the way she had empathised and twice dragged out his name. 'Follow my wife, Mr Ross. She knows all the short cuts through this blinking kafuffle.'

'What did Bjorn just say, Kirsten? That's new to me. Was it Australian slang?'

'Forget it, Gigi. It's his wacky Aussie lingo. It's a wonder he hasn't called Ian mate.' Whilst speaking, Kirsten walked in front of the men to their car. Instead, Ian walked towards a sign marked: "Heathrow's main carpark".

'Gigi, is Ian always this surly?' Kirsty was mystified by his arrogance. 'I don't mean to be rude. It seems to me as if he's too absorbed with his own importance to be polite. Hope it's not because I insisted on you both staying with us overnight?'

'No Kirsten, he's tired. Ian received horrible news before we left Zurich.' Gigi changed her small case to the other hand. Numb fingers restricted her holding it any longer. 'His aged mother is ailing. He'll be flying to Scotland, after I leave for Llangollen tomorrow. In the morning please remind me to verify my flight time to Manchester.'

'Bjorn will do it for you first thing.' Kirsty paused. *In the morning I'll tell her about our news.* 'If Ian doesn't want to stay with us, there's my neighbour's flat. Evika won't mind putting him up for one night. You do remember Mrs Patchcoick?'

'Yes, I recall the antiquated way she spoke in a Russian dialect. Don't worry Kirsty, we can sleep in the spare bedroom…if you're short of space?'

'No darling, you're sleeping in my bed…with me. The men can bunk down in our guest room. You must tell me what sort of a lover your man is, Gigi.'

From the look she received, Kirsty knew this was a touchy subject. She responded haughtily. 'Oh well, girls will be girls! And all men think they're Casanovas, or mine does.' Inwardly she chuckled. *Bjorn's harmless. Though I did notice Ian seems to have a roving eye for the ladies.*

'After four years of wedded bliss, you're still very much in love with Bjorn. I hope our marriage will be as good as yours and BJ's…'

'Of course, we still love one another,' Kirsten interjected setting her guest's hatbox and make-up bag on the pavement. She decided against commenting on their forthcoming marriage and their future. 'Gigi, you're the only person who calls him BJ. He loves you calling him that. It reminds him of Australia. Quite homesick is our boy.'

'Have you heard any more about your quest'? Eager to hear her reply, Gigi almost tripped over a stranger's cumbersome suitcase and swore under her breath.

'Yes, we leave London in four weeks. Bjorn's had enough of working in the airline business. He's decided it's time we headed home to Australia.'

'Nothing would surprise me where your husband is concerned. Tell me more, I'm all ears Kirsten.'

Gigi knew Flight Captain Svensson was restless and need to leave this job. Bjorn reckoned flying was monotonous after a ten year stint. Constantly travelling as a "sky-driver" wasn't his idea of pleasant work. Not when there were more positive options in the wind.

'Guess what? We've bought a sheep station. You know, baa-baa's.' Kirsten placed a thumb on each temple and wiggled her fingers, indicating horns. 'It's his uncle's stud and we'll make it a going concern. Not a cattle stud. Though there are several hundred Herefords on the property. They're killers for our home use only.'

'Where's your property in Australia?' Gigi gasped as Kirsten whispered it and gave her a superfluous grin. 'This news of yours is terrific. You originally came from that town, didn't you Kirsty? I'll come and plague you with my company, once you move in. Australia's a massive country, going by this brochure I found at the airport.'

Kirsten nodded in agreement. 'Here's our car now. I thought Bjorn might've let Ian drive. Instead, he's changed his mind. I'll tell you more of our adventures tonight in bed. Those two worriers won't be snooping or annoying us in there. We can natter to our heart's content without being disturbed.'

Once the Peugeot arrived at its destination Ian Ross refused to alight. *I'm not their slave. They can unload this damn antiquated contraption without me.* In the foulest of moods, he scowled at Gigi whose harsh frown showed her anger. He knew the glower was in retaliation for his refusal to help Bjorn with their cumbersome luggage.

Reciprocating with a despicable snarl he ignored her pout. 'Staying with *your* friends wasn't my idea. You had a hide accepting their offer, without first consulting me.'

When we're alone I'll tackle her about this inconvenient arrangement. Exasperated, he thought: *We can book into a hotel. Somewhere quiet where we can be alone without people intruding on our privacy. If it's a hovel, at least we will have the freedom to pursue a passionate love affair. I hate people who can't mind their own damn business.*

Mentally reflecting on their confrontation Ian thought on entering the Svensson's small flatette. *We'll be going our separate ways tomorrow. Walking through London alone to find somewhere to bed down overnight, while she'll be enjoying herself in Wales will be pure unadulterated hell for me.*

The romantic evening Ian had planned now lay in tatters. Forced into a corner without an escape was inexcusable. It reminded him of the darkest period of his life in Germany, when the country was threatened by battle-weary Russians who pillaged, raped women and destroyed his beloved Fatherland.

Confronted with sleeping in the guest room, he bluntly refused.

'Even the idea of bunking down with that self-opinionated idiot repels me. Give me a good reason why can't we leave here tonight? Otherwise those nosey friends of yours will keep interfering in our lives. You know I'm damn well right, Gigi.'

She refused to insult her friends by leaving and it caused another row. In conflict, they resorted to resolving their differences in Kirsten's car. This allowed them to retain their privacy without interference. Tempers at flashpoint spoilt her evening. The loser of their row, Ian relented. In a huff, he agreed to doss down on the lounge.

This suited Bjorn, who'd tolerated enough of their arrogant guest's rudeness.

'It won't hurt the ignorant snob to sleep on that old lounge, Kirsty. With a bit of luck, its crooked springs might buck him to hell before morning. It might force the bastard to consider us *and* Gigi. All he thinks of…is his own sanctimonious self.'

On retiring, Gigi thought Ian's belligerent attitude was uncalled for and she resented it and him. She swore never to return from Llangollen in Wales, if his temperamental bouts of anger didn't subside. She also resented the despicable manner he treated and spoke to her. Especially how he'd threatened to leave Heathrow Airport without saying goodbye.

Having overheard them arguing in the car, Bjorn returned indoors and spoke to his wife. 'If she's mad enough, Gigi will make that arrogant prick think of leaving here earlier than expected. It can't happen quick enough for me.'

Ian's idea was to make her to believe in his sincerity. He couldn't concentrate this evening, because of her obstinacy and stress. *How could that Welsh bitch have penetrated my defensive shield? I studied the psychological mannerisms of devious prisoners at the Chancellery, so why should I tolerate her belligerence now? She will never become my nemesis. I will be the dominant force in my household. Try as she might, Gigi will never control me... or my mind.*

Until midnight the women talked on various topics including their brief sojourn with the Sadler's Wells Ballet Company in London, where their close relationship flourished. They worked together in Paris under Monsieur Bouvier's tutorship for five years. Pierre later became Gigi's personal coach until her and Kirsten's decision to go their separate ways.

A wealthy Parisian entrepreneur, he considered this Welsh dancer his dearest friend. He respected Gigi and classed her as a close member of his family. As her benefactor, Pierre had covered all her medical bills in Switzerland. And as a consolation he offered her a partnership in his ballet company. A gratuitous feeling induced her to think passionately of this kind and gentle Frenchman whom she could never repay in a lifetime.

After tonight's violent argument, she thought Ian's slanderous defamation of Pierre was barbaric. Unable to dismiss his rudeness, she cried until sleep claimed her mind. In her dreams, she could still hear Ian saying, "If you refuse my love for his, I'll find another woman with whom to share my life". Now those words had the potential to become the focal point of their ongoing discord. This she regretted. Powerless to reverse her condition, she feared telling Ian of her suspected pregnancy.

'Hope lingers on the horizon, if I don't antagonise him. I will be careful what I say in future. I shan't tell him now about our babe, whom I might name Kendall.'

5

6 am. Kirsten, who received a telegram last the previous night, gave it to Gigi at the breakfast table. 'Marked urgent, I opened the envelope and read its text, Gigi.'

She glanced at the words which read: "Gigi, Manny hasn't long to live. Her heart is failing. I think you should be here. Please come home. Your cousin Jen."

Having discussed the gram with his wife, Bjorn had managed procure a seat for Gigi on a commercial flight to Manchester at noon today from Heathrow. Their decision to keep this news from her was sensible.

Bjorn didn't tell Gigi that he'd found an empty whiskey bottle under the lounge. Unable to arouse Ian, he thought it advisable to let him sleep, until after breakfast.

8 am: In the shower Gigi was confronted by Ian, still under the influence who tried to seduce her. One whiff of his breath made her heave. 'No, don't come near me until you brush your teeth. You stink of booze and your breath reeks of stale tobacco. Ian, you shouldn't walk around their flat in the nude, either. Please have some respect for my friends, if not for me. Go and put on some clothes.' Disgusted with his crude antics, she dressed and left the bathroom to seek solace in peaceful surroundings of Kirsten's room.

The thought of food repulsed him. He refused to eat. Ian felt his head would burst from anger, and he challenged Gigi in the hall.

'Why must you keep insisting that we stay here, when you know that smart-arsed Svensson despises me?' Ian grasped her wet hair and twisted it until her eyes focused on his bloodshot eyes. 'It's not before time you took notice of what I say, instead of listening to those idiots. They don't care about you. Wake up to yourself and stop walking around with your nose in the air.'

His backhanded thump collected her jaw which sent her flying against the wall. In a crumpled heap on the bathroom floor she wailed piteously. With the same fist raised, Ian stood over her ready to bash her head on the marble washbasin, until he heard light footfalls approaching from the hall.

Kirsten changed her mind from entering the toilet as Bjorn pointed to their lounge room.

'Darling, if I don't arouse that smug bastard, Gigi could miss her flight.'

'Give him a few more minutes. Oh damn, there's the phone. I'll answer it, while you have a shower. Then you can wake him.'

As their footsteps faded, Ian crept out to the lounge room and collected his clothes. In the guest bedroom he dressed then went to the bathroom to shave. Gigi wasn't there.

'Ha, she'll keep until we're alone at Heathrow. As that mirror is my judge I'll make her pay for siding with those ill-mannered bastards.'

When time permitted and conditions prevailed, Kirsten intended to have a good talk to her. She knew Gigi would ruin her life by marrying Ian. Although she couldn't define her strong resentment of him. There was something ominous about his brusque manner which puzzled her. *When he talks to Gigi, his brogue doesn't sound Scottish. Mmmm, it sounds different to the way Jock MacDougal speaks to his wife.*

Bjorn had drawn the same conclusion.

'Don't worry Kirsty. If need be, I'll phone Pierre in Paris. He won't hesitate to contact the Chief Constable here. With his sway, the London bobbies will soon send the smart-arse packing for good.'

Unbeknown to Bjorn, his wife had spoken to Gigi in their bathroom. Tearful, and while bathing her bruised face, she pleaded with Kirsten not to divulge their row or Ian's irascible temper to their French mentor.

'The dear man, he has enough worry with his wife ill. Kirsty, please don't phone Pierre, or his family. I don't want them knowing out about our engagement. I'll tell him myself when we're settled. Everything will be fine once Manny's well again. My cousins live in the same street and will take care of her needs. My mind is made up. I do know what is best for us. No one will prevent me from marrying Ian. Not you or even Pierre.'

Gigi promised to keep the Svenssons informed of their wedding plans. Until then, she and Ian must be allowed to decide for themselves.

She adored him and intended to have a fulfilled life with him, without anyone's interference.

Ian's moroseness worsened by the minute. He remained silent when Gigi mentioned her early departure. *Her altered plans conflict with mine. I'll talk her out of siding with them. It won't be easy. The Svenssons are determined to separate us. I know the bastards are trying to persuade Gigi to break our engagement and they want her stay in Llangollen.*

In a fit of anger Ian threw his breakfast in the kitchen bin. He then cornered Gigi in the bedroom.

'Gigi, we need to discuss a few important things before we leave here. Let's go for a walk. The Svenssons have your welfare at heart, and I know they dislike me. Fetch your hat and coat. I'll wait for you downstairs, near their car.'

'No Ian, I can't afford to miss my flight. You know it leaves Heathrow at noon.' Her terse response hit the mark. Gigi knew if she relented, he would keep harping until they reached the airport. Collecting her coat and a scarf from the hallstand, she followed Ian downstairs.

She tried to explain that leaving a day earlier shouldn't matter. Manny needed her in Wales and he must accept her decision. Her distant cousin would be waiting to collect her at the Manchester Airport.

'Ian, I expect you to understand my situation. I know of your predicament with your mother being so ill. Please try to be more tolerant with both me and the Svenssons.'

From his bedroom balcony, Bjorn observed their guests strolling down the road. 'Neither of them know their way around this district. Ian will deliberately detain Gigi, to prevent her leaving London. Rows of the conjoined tenements in these outlying suburbs all look alike.'

Bjorn spoke to his wife while retying his shoelace.

'Nothing would surprise me about that ignorant deviate. He has a bloody lot to learn about how to treat women with respect. Kirsty, you heard what went on in the bathroom this morning, or part of their argument. My heart aches for Gigi, especially if she does marry that swine. He's a gutter worm from the sewers as far as I'm concerned. If he ever bashes or harms her in any way, I'll kill the smug bastard.'

'Darling, she must make up her own mind about their wedding. I tried to talk to her this morning. It was useless. Gigi insists on marrying him. Ian won't let her out of his sight until we reach Heathrow.'

He forced her to walk one step ahead of him as they ploughed through slush. Then he abruptly stopped. His glum expression and his balled fist sat an inch from her face. The frightening glare of anger in his eyes made Gigi edge away from him.

'What have I done now, Ian? You're scaring me.' As she stepped off the wet pavement his hand steadied her. Together they crossed over to the grass verge.

'Let's sit here, on this wooden bench, it's dry. If you're worried over missing your flight, you won't.' His tone mellowed and he encouraged her to smile by raising her fingers to his lips in an endearing caress. 'Darling, what I have to tell you *is* extremely important. I need to say what's on my mind, before we go leave today, at noon.'

Her gasp to his 'don't interrupt me look' frightened Gigi.

'I refuse to let anything or anyone come between us, Ian. My love for you is genuine. I won't argue with you. Nor do I wish to be separated from you, longer than necessary.'

'Do you *really* love and respect me? I doubt it. Not after hearing you and Kirsten discussing Pierre. I *will not* put up with your sullenness. I should be the *only* man in your life. Not him. Make up your mind now. If it's him you want, I won't be here when you return from Llangollen.' She knew by the harshness of his voice that Ian meant every word. 'You must decide now. Or you can find your own way back to their damn flat. I never want to hear you say that Frenchman's name again. Do you understand me, Gigi?' Wrenching her arm, Ian forced it back. 'Your constant prattle of him is enough to drive a sane man crazy.'

'Stop it. You're hurting me Ian,' she pleaded. 'Look, you've left a bruise on my wrist. Leave me alone.' Angrily she pulled away. 'How many times do you want me to say, I love you? I want and need you. Not Pierre,' she whimpered as the frantic tone in her voice heightened. 'You know I wouldn't be going to Llangollen if it wasn't urgent. You'll be home with your ill mother in Inverness. I'm not changing my plans, on a whim of yours. Forget it.'

'Will you cease arguing?' Ian took a small parcel from his pocket and passed it to her. 'Don't open this gift until you're on the plane. Better still, leave it until tonight.'

Misty-eyed, she found it difficult to focus on the brown package. Shaking, her fingers gripped the gift. Remorseful and feeling guilty, a spasm of pain struck her heart, which at present ruled her head. She knew this constant chatter of her ex-dancing partner was the antagonising factor in the lives, which inflamed a delicate situation. *I'm wrong. Ian was right to forbid me from talking about Pierre. Perhaps he thinks Pierre will be a threat to our marriage?*

Embarrassed and ashamed, a demure look overshadowed her scowl.

'Darling, let's go back to their flatette. Bjorn hates driving fast and being caught in heavy traffic. All the main roads to Heathrow will be busy this morning.'

'Not yet. You must be aware how I detest their arrogance. They disbelieved my promise to take care of you once we're married. It's belittling to have my authority disputed and overruled by the Svenssons,' Ian snarled. 'You constantly defy my request not to flaunt that Frenchman's name in my face and it's annoying. Promise me not to ever mention him again *Schatzi...*' *Hell,* he thought ... *I've done it this time.*

The startled look on his face imaged hers.

'What did you just call me? It sounded like Nazi or nasty. Or was it a foreign name.'

'No, you're wrong. I meant the tiny flowers in the church garden we just past.'

Mystified, she couldn't grasp what he meant, until it clicked. 'Ah, the pansies. Ian, my dearest wish is to be your wife. I'm sorry. It was wrong of me to mistrust you. The next four or five weeks will feel like an eternity to me in Wales.'

'You're already mine,' he scoffed. 'We've consummated our love. Remember this Gigi, no man on this earth will sexually arouse, or touch you in a familiar way again. Not while I'm alive.'

'You made that sound like a...' She hesitated, fearing to say a threat. Let alone say what her mind conjured. *No, I'm imagining what he meant.*

'A threat! Yes. Now you know how your sarcastic comments hurt me.' His frown diminished and he turned to face her. 'Gigi, I am a jealous man. You wouldn't have me any other way. Not if you're being honest with yourself.'

Ian caressed her and their lips melded in love. Then as her hand sneaked beneath his tweed overcoat it heightened his sexual drive. His smile reflected in her eyes. Convinced he had successfully bought her silence and undying love with a small gift, he accepted her apology. Ian found it gratifying to think how a trinket could make her obedient to his demands. *The power of logical thinking has swayed her infantile mind never again to speak of the man whom I detest.*

Gigi began to panic. She knew if they didn't hurry she would miss her flight to Manchester. She accepted Ian's assistance to step over heaps of debris. A legacy of war!

'Do *not* say what I've discussed with you to the Svenssons. Nor speak of the pact we've made today. I couldn't let you walk out of my life for weeks, without bringing their belligerence to our wedding and future out in the open. Their blatant attitudes are intolerable. They belittle you in every way possible. Gigi, don't ever deceive me. If you do, you will regret lying to me.'

'Does that mean you don't want me to phone *him* or his family?' She almost said Pierre. 'Ian, if I don't keep in contact with them, they'll think something's wrong with me ...or my health.'

'The answer, I'll leave in abeyance until you return from Wales. We all need friends. I couldn't do without mine. Darling, you mustn't take everything they tell you to heart. While you're in Llangollen think of us being together in our new home. It'll boost your sombre moods. I know you'll feel as miserable in Wales, as I will in Inverness.'

Strolling past a Catholic church Gigi was tempted to enter the vestry. *No, there isn't time. Our wedding will be in a small church. I'll be on tenterhooks until we say our nuptials with friends to witness our vows. All that matters now is Ian has agreed to let me leave we've settled our differences.* Holding his hand, she looked dreamily into his eyes and could feel pride swelling in her bosom as they hastily returned to the Svensson's flatette in Highgate Wood.

Ian repudiated the suggestion for him to leave for Inverness without arranging a point of contact for Gigi to access in an emergency. *I can't afford to have another row with her...or them. Not with her trust of me restored and she's agreed to my demands. Any disruption will weaken the power which I now wield over her. I know if she's alone with that witch of a bitch, Gigi will again accept her advice to leave me. Then she'll disrupt my plans and refuse to return to London.*

Packing the last of his clothes Ian smiled. *Finally my strategies are beginning to slot into place. By the time I finish with the Svenssons and that French smart-arse, they'll regret encouraging Gigi not to return from Wales. When we're married I'll demand her obedience and her allegiance to me, only. My stratagem of keeping her isolated will make her subservient to my every command. They all think I'm a dim-witted fool. If I don't offer to control her financial difficulties, she'll repudiate every proposal I put forward. Perhaps then, she'll rely on the foolish ideas they've implanted in that juvenile mind of hers. My greatest challenge will be if she rebels and turns against me. If I convince her that our future together will be utopia, she'll relent. In a year or so, she'll be my ticket to freedom. Away from these rabble-rousing idiots.*

Looking at his reflection in the bathroom mirror Ian scowled.

'It's humiliating the way she toddies to that French bastard's every whim. I must persuade her to dismiss the lies he and this lot of fools have drummed into her stubborn head. Or there's a possibility of her rejecting me and my love. Gigi repudiates everything I say and then she's angry, or unbelievably disagreeable.'

Ian knew by constantly harping of his dislike for the Svenssons and Pierre Bouvier, Bjorn would try to encourage Gigi to break off their engagement. His hate, governed by dissention, was growing stronger for this trio of snoops. His anger festered because they were manipulating her life. Thinking in a more positive vein, his mind functioned like a well-sprung clock. He could churn the wheels of fate to suit his plans. Not hers. *I will win this battle of wits. I've sensed a wall of dissension building between us for a while now. A catastrophe must be averted, until we part company at eleven this morning. Controlling my temper is important. I'll try to be courteous and casual until we reach Heathrow.*

'Gigi darling, I *am* going to the airport with you. If all goes well, I'll book in to a hotel somewhere close. Then I won't have to worry about transport.'

'Ian, I took it for granted you'd be coming with us to Heathrow. Why the change of mind? Yesterday you promised to stay with the Svenssons until tomorrow. Be truthful! Aren't my friends good enough for you?'

'No, you're wrong. The truth is…I have an important appointment, an interview in London. In my line of work difficulties occur daily. My business affairs there are urgent.' Assisting her down the lounge room step, Ian concluded, 'They're matters I should've attended to in Zurich. Enthralled with the prospects of marrying you … I forgot. My business in town should be finished by twelve. Your flight leaves when?'

'Right on noon,' she responded, disillusioned with his gruffness and trite manner. 'I hope Manny is okay until I reach home. Her health is failing. One is never sure when life will end. You know it's true, Ian.'

'Life's span begins the moment of conception. It wends its way through the years until we're elderly. My dear, I wouldn't want mine snuffed short of forty years. Further on our travels together, I'll accept beleaguer of death with you by my side.'

The realisation of how many lives had been expendable never perturbed this ex-Nazi officer in the slightest. All criminal episodes belonged to the past. There they would remain. Ian recalled the topic he'd discussed with Gigi before leaving the Edelweiss Pension in Switzerland. It related to children. She desired *four*. The *word* rested uneasy on his mind. *What, four little ankle-biters? Hell, I can't think of anything worse than another tribe of squawking brats underfoot.*

While washing his hands in their bathroom, Ian recalled her words about children and shuddered. *Those brats of mine were damn nuisances. They're beyond the River Styx now. There they can neither harm us, nor incriminate me in their mother's death.* The cruel gleam in his eyes and distorted lips imaged an evil creature in books or on film screens of the red horned beast who stoked the fires of hell. Adjusting his cinnamon tie under the collar of his steel-grey suitcoat, he smiled. *Yes, and hell is welcome to them.*

On their arrival at the Svensson's flat, Ian had refused to remove his luggage from the car. Leaning over the balcony-rail Gigi caught a glimpse of Bjorn scurrying indoors. She naturally thought he may have gone down to speak with Ian.

Instead she discovered him locking his leather valise. She glowered at the forbidden briefcase's eagle.

'Why aren't you ready, Ian? Bjorn and Kirsten are ready to leave. If you dawdle here much longer, I *will* miss my flight home.'

Kirsten eased past Gigi in the bedroom.

'Mr Ross, you're welcome to stay here.'

He thumbed her. 'I have important business in London. At Heathrow, I'll wait until her flight is ready to leave. Then I intend staying in the city. So you, Mrs Svensson and your husband needn't bother hanging around to wish Gigi farewell. We need some time together, without inquisitive people snooping into our privacy.'

His crass remarks stunned Kirsten. She turned and followed Gigi down to the car.

'Tough luck, old fart,' Bjorn smirked blocking his way outside the spare bedroom door. 'You'll find it hard to find digs in the city, now matie. Everything's booked solid.'

'Move out of my way, Svensson. I'm not worried. Money counts, or haven't you heard? Something will turn up. It always does. Don't matie ... or old "fart" me. Either you refer to me by name, or not at all.'

Bjorn's contemptuous snigger didn't go down well with his parting guest. 'Look, I know we started on the wrong foot. Please give me a chance to apologise, Ian? After all, we both want what's best for Gigi.'

'How can I finish dressing, if you persist on blocking my way,' Ian snarled. 'Tell Gigi she better not leave without me, Svensssson,' he hissed. 'Now, step aside and allow me to enter this room.' Brushing past Bjorn, he almost sent him flying against the mahogany hallstand.

Bjorn didn't bother retaliating, he just scowled at Ian. *I feel like wiping that supercilious smirk off his dial and smashing the idiot's skull against that wall.* Instead, he curbed both his tongue and temper, to a degree.

'We'll be leaving here in five minutes. If you're not ready Mr Ross, it will be my pleasure to call a cab. Then you can collect your cases from the gutter. It won't mind your *company*, whereas I find *it* and *you* repulsive. Remember, five minutes, not a fraction longer.' Fuming, he slammed the bedroom door and almost collided with his wife in the hall.

'Did he ruffle your feathers, Bjorn? Your face is scarlet.'

'Why wouldn't it be? I almost kicked that arrogant bum where it hurts the most. Thank goodness he's staying in town until his flight leaves. One more second with that inconsiderate bastard and I'd have decked him flatter than a tack.'

'Darling, he'll find it hard to...'

'Kirsty, I warned him. Nobody can penetrate that stinking drunk's rhino-hide. Don't know what he does for a crust. I pity the poor bastard who he works for. He thinks his...'

'Ah ah, don't you say it Bjorn. Bull-dust...if you must say something.'

'That smug-arsed bastard gives me the irites. I'll go and see if Gigi's left anything in your room. Don't say a word to her about our argument, or how much that miserable tightwad annoys me.'

As he began to walk away Kirsten nudged his elbow.

'I have the solution to this problem. We can't really put my idea into practice until we've been home for a month or two. I'll talk it through tonight, darls. Then, I'm going to break a promise.'

'Struth, that sounds ominous, Kirsty.'

Gigi appeared in the doorway, holding her hair in place. She frowned.

'What's this about a promise? I didn't mean to intrude. I only came to borrow your comb. Mine's packed.'

Kirsten said the first thing that popped into her mind.

'Bjorn's offered to take me to see Swan Lake in town next week. I'll book our seats then for Puccini's La Bohème. Its season begins soon at the Theatre Royal in London.'

Gigi caught on. She'd seen the operas advertised in the *London Times*. She thought to herself, *knowing she's a theatre buff, Kirsten would dread missing a Command Performance. I wish we could go with them. Ian hates operas. That means I won't be going to any theatre reviews here. I shan't be home from Wales until later in the month, if then. I must remind Ian to give me a contact number, before I leave Heathrow in two hours.*

Fearing he would explode with laughter, Bjorn pushed past his wife and whispered 'Liar,' then hurriedly vacated the hall. Combing his hair near the front door, his smirk developed into a full-blown grin. *That idea of my wife's pre-empted mine. I better make a booking for that performance, or I'll never hear the end of it. I'm curious to hear what her proposal is for Gigi. Whatever Kirsty's idea is for their future, it can't eventuate until we're back home in Australia. She has a knack of solving difficult problems, does my wife.* Bjorn didn't think Gigi had the time nor could she wait.

'God help her if she ever has kids to that bastard.' This hushed refrain issued forth unintentionally. Turning he saw Ian standing directly behind him. *Too bad if the bastard heard me. It might make the bastard see reason.* Bjorn laughed collecting his overcoat and hat off the umbrella stand.

'Where's my damn keys, Kirsty?'

'Already in the car with your suitcoat and we're ready to leave. We'll wait for you and Ian downstairs. It's time we left for Heathrow. Don't be long, Bjorn.'

As Gigi's flight soared skywards Kirsty turned to say something to Ian Ross. She was stunned. There was no a sign of him, or his hand luggage anywhere.

'You know darls, he didn't even have the decency to say goodbye.'

'What do you expect from a smug ignoramus? I didn't expect him to fall over us with platitudes. Now you know why I threatened to deck the bastard this morning after he tried to goad Gigi into another violent argument.'

On the way home Kirsten discussed her idea with her husband.

'What evil scheme is that devious man concocting now? He's been trying to stir-up trouble between us and Gigi since he arrived here. The suspicious look he gave me, before he left the airport, indicated that he mistrusts us.'

'Little Miss Innocence can't see that his deviousness is making her feel the guilty one. Gigi thinks the sun shines out of his arse. Unfortunately, that sanctimonious creep is blameless where she's concerned. I wouldn't trust Ian any more than I can lift one of our draft-horses.' Listening to his wife's proposal, Bjorn agreed with her proposal. 'I think you've square-pegged their controversy, Kirsty. Now I need time to think it over. We'll discuss it later tonight in bed. Did you give her our address in Australia?'

'How could I, Bjorn? Once the final papers for "Jalna" are signed, I'll either phone Pierre, or jot a line to her in Llangollen. He keeps in touch with her.' Approaching their nearside kerb Kirsty uttered a sigh. 'If Manny's health deteriorates, I asked Gigi to give us a call. Then Ian won't know she's contacted us.'

'Smart thinking.' He nodded in agreement. 'Did you think to tell her to reverse the charges? She can do the same once we're settled in Yass. It'll be six to eight weeks before we can take possession of the property in the Southern Highlands of New South Wales. That's if this four-leaf clover works its magic.' Bjorn flicked the oval-amber talisman on a gold chain around his neck. 'The trouble is how to confront Ian with your proposal. He sure as hell will buck, if Gigi's accepts it without telling him.'

'No darls. He won't be game to block our proposal. Not if he wants to marry her. I spoke to Gigi in our room, just before we left Highgate. She does have some knowledge of what he's planned in London. Still, it will be interesting to hear how everything works out. In time, or on her return from Wales she may confide in me.' Bjorn just smiled.

6

July 1946 in Wales: Gigi couldn't contain her frustration. Overanxious from not hearing from Ian, she expected to receive a letter or a telephone call from him. Not even a postcard had arrived in Llangollen and she began to suspect he was carrying on with another woman. After three weeks of loneliness she prayed for him to phone. When he hadn't rung by nightfall on the fourth Wednesday, her moroseness increased fourfold and she cried herself to sleep, as she did most nights. Sometimes in front of a struggling fire, due to the inclement weather. Downpours were a daily occurrence and the little wood in the firebox was damp. Storms also hindered her from collecting dry bits of kindling from under the drenched woodpile.

When Ian finally condescended to phone his platitudes were less than pathetic. She listened as he instructed her not to return to London.

'Gigi, I'm staying at the Dorchester Hotel in Gloucester. It's in the Cotswold Hills. You have map, use it.'

In anger, she responded in a belligerent manner.

'Ian, you know it's impossible for me to be down there by then. From Llangollen it's hard to travel anywhere. I'll try to arrange something this end. I can't give you an answer or leave here until after my grandmother's funeral tomorrow.'

'Fair enough. I'll ring again in two days. Figure out something positive before then. Or you'll have to make your own arrangements from London to here.' No address. No love. No nothing! Imparting an undignified growl, Ian slammed the receiver down, which left her a quandary.

Apart from being hurt, she felt sick.

'There's a slim chance I can phone my cousin in town. Why must Ian be so demanding?' She dialled a local number and left a message. An hour later he verified that he and his wife would be leaving for Oxford on Friday and offered to collect Gigi on their way down through Gloucester then to Somerset.

With no other choice than to wait until Ian rang again, she began to worry.

'Why didn't he give me a phone number? He's never punctual, especially after promising to call me. Surely he's settled somewhere by now.' She scowled, throwing a matted, blue slipper at the bedroom mirror.

Distracted by reminiscing, the steak had burnt to a cinder. In a dither, she flicked off the stove. Burning two fingers, her elbow nudged the hot pan over to the cool hob.

'Dash it, I haven't any more meat. A cheese sandwich will do. This nauseous feeling keeps making me heave. Even the idea of food makes me want to vomit.'

Every time the phone rang she began to get excited. Most incoming calls were from friends as condolences. Reduced to tears, she tried to contact the Svenssons in London. They were in Australia. Gigi had missed them by four days. In desperation, she tried to ring Paris. The lines there were congested. *If anyone knows their new phone number, Pierre will. I'll try to ring him again in an hour.*

After extensive hassling and loss of patience and lack of food, she managed to reach his home number in Montmartre, a Parisienne suburb.

With Pierre away Raoul, his adult son answered the phone. Gigi briefly outlined the reason for her call, to which he responded, 'I'm here to collect a few items I need for my class at the Conservatorium. I'll be leaving in ten minutes, Gigi. I understand and I'll leave a message for Dad to ring you the instant he comes home.' The young musician sensed her plight, and asked if she was ill. 'Your voice sounds strained, are you feeling well? Or is anything wrong, Gigi?

'No, I'm fine Raoul. Darling, I urgently need Kirsten's number in Australia. Please try to contact your father? Pierre knows my number here. I'll be leaving Llangollen on Friday. God Willing, nothing else will go wrong before then.'

'Remember, you've precious to us. Please take care of yourself climbing in those bleak Cambrian Mountains.' Raoul glanced at the library clock.

'Sorry, I must fly. I'm jotting down a note to Dad. Then I'll leave it on his desk. Bye sweetie.'

In Llangollen this despondent young woman lay on her bed and cried inconsolably. Scared and lonely she desperately wanted someone close to console her.

'Autumn has relented to its icy cloak to winter,' she gasped and shivered. Deluging rain of the raging storm swept in every crevice of this old home until its floorboards creaked. The creaking trees, their boughs heavily laden with snow frightened her. Gusts of wind howled through the thatched, pitch-sealed roof, now sodden with rain. Growling thunder sounded like someone whiplashing the obsidian sky. The house walls began to shake and blinding flashes of lightning made her feel even more despondent.

Consolation came in the form of a letter, photos and negatives from Ruffina, which she received by mail that morning.

'What use are they now? Ian doesn't care if I'm dead or alive,' she wailed, perched on a chair in the freezing lounge. The chairs and sofa were saturated. Water seeping through the ceiling dripped in buckets. Tempted to destroy the miniatures, she relit the fire with damp wood and scraps of newspaper, as an idea formed in her mind. *If I hide these photos and negatives in my curler-bag, Ian will never think of looking there.* Blubbering and in a rebellious mood, she didn't hear the phone. It buzzed for a while before she realised.

'Is that you, Pierre?' Gigi queried, sniffling. Leaning her head on a tear-stained cushion she listened for a reply.

'No, Raoul speaking. Gigi, I managed to locate Dad. The Svensson's number in Australia is a new connection. Hurry and jot it down, or I *will* be late for my music exams at the Conservatorium.'

Gigi wrote the number in her diary and thanked her godson. In a fortuitous move she managed within minutes to get a direct line through to Australia.

'Kirsten, it's me, 'the excitement in her voice reached fever-pitch. 'I'm leaving here on Friday.' The intermittent line sounded hollow. 'No, I'm not going to London.' The line again buzzed with interference. As it cleared she sighed. 'Ian wants me to go straight to Gloucester. You know how secretive he can be, and I don't know why.' Static on the line made it impossible to hear. She paused until it cleared.

'Canberra is the capital? Your property of Jalna is how far from there? More than I expected. I'll be anxious to see your photographs. Yes, I do miss Manny terribly. I held her hands and bathed her throbbing head as she said her last words to me before…'

A rumbling noise distorted the line and she cursed. Gigi relayed some details of her wedding plans. 'I'm sorry you and Bjorn can't come to England. I understand you're both busy and it's a long way to travel.' The line dropped out. Disappointed, she wanted to wish the Svenssons all the best for their new enterprise, a wool growing business and a large self-supporting homestead in Yass, New South Wales.

'Kirsten will understand,' she moaned, relighting four candles in the cold kitchen. 'Damn this wretched storm. When will it cease? I'll put another saucepan in the hallway to catch those never-ending drips.' Shivering, she pulled Manny's hand-loomed woollen shawl around her shoulders. The wind increased and thunder grumbled close by. Gigi shuddered as the roof groaned and a flood of disgorged water saturated her. Twisted flames licked wooden logs, as she endeavoured to keep warm by the hearth. A pitiful glow struggled to survive above the partly-demolished brick chimney. Bending over the hearth, a clump of loose soot doused her hair.

Again she cursed, while trying to rekindle the fire.

'I wish Kirsten could be here with me. She'd never be afraid of these harsh bursts of thunder, or sheet and jagged lightning. Unlike me. This old house is creepy, especially when wind rattles the cracked windows.' She ceased grumbling to move a pot of vegetable stew over to the stove-hob. The coals of her established fire glowed red. 'Jalna sounds huge.' Warming herself by the hearth she thought: *I'm sure Bjorn and Kirsten will make this venture work to their advantage. Yass sounds a beautiful, yet small town with its surrounding massive acreages open to the sun-bronzed sky.*

The sudden realisation that she had more to worry over than a sheep property made her embrace the washbasin. Vomiting until her stomach ached and disillusioned after hearing why the Svenssons couldn't attend her wedding didn't improve her melancholia. Nor did it enhance her glumness. And it certainly didn't ease this severe bout of nausea. Depressed, the threat of taking her life lay constantly on her mind. Reared in the Catholic faith she dismissed the idea of suicide. It was a cardinal sin and against the church's principles. Confused why Ian hadn't phoned, Gigi washed down a prescribed pill with a swig of vile-tasting tank water. Feeling miserable she sank on her grandmother's bed and cried, until sleep dulled an overactive mind.

5 am: Disturbed by the cry of fox in the timberland, she awoke and sized up her life. How everything had reverberated with a pendulous swing of indecisive reflection on their relationship. In four weeks, the lows had outweighed the highs in this wintery world of uncertainty. Her life was diverting along the unnatural path of illness accompanied by loneliness. Deception by Ian, coupled with increasing sadness of her Gran's death, drove her to the brink of insanity. Isolated in this derelict Celtic cottage with rain deluging its saturated thatched roof, she began to wonder if it would ever cease. Her grandmother's gnarled pear tree's twisted boughs thrashed against each miniature piece of triangular glass in the windowpanes.

'Why has God forsaken me?' she wailed, trying to cope as lightning emblazoned the pitch of night, amid a power outage. 'Where's my pocket torch? Oh yes, it's on the washstand in my room.' A lapse in massive claps of thunder, she hurried down the narrow hall to retrieve her torch and a dry box of matches. The candle-holder rattled. She cringed in fear as another ear-shattering clap of thunder shook the house. Carefully she detoured around pots full of rainwater, as more seeped through the wet ceiling. On her return from Manny's room, she saw letters being poked through the brass slit in the front door. Quickly she rushed to snatch them before they fell on to the wet floor, Gigi opened the door. Relieved it wasn't kids playing pranks, she smiled at her grandmother's postman.

'Sorry I couldn't deliver this bundle of letters on my last round yesterday afternoon. We all miss Manny. Now she's at peace. There's a special letter in that pile marked for you, Gigi. Are you going to sell her home? If so, I want first chance to buy this old house. Name your price and a cheque will be in your bank within a week. Have a better day, my dear.' His frown was one of concern. 'You look a little tired, dear. Give me a buzz on the phone, if you need help. Now I need to finish these deliveries before it rains again.'

'Thanks, Mr Colwyn. I'll arrange with my lawyer to sell you this house. It might take time. You have my word that I won't sell it anyone else. Give my regards to your wife.' She waved as the elderly man climbed back on his pushbike.

Sitting in the icy parlor Gigi read *her* letter.

'Well, at last he's letting me know the drop-off point near his hotel. Ian's quite lackadaisical at times. I'm sure my cousin will know the nearest spot to set me down in Curzon Street. Wish my knowledge of that part of England wasn't so sparse.' With time to spare before her cousin's arrival, Gigi wrapped Manny's three miniature porcelain figurines of an old-world couple in a shawl with her jewellery. *This bundle should be safe tucked in among my winter sweaters.* Locking the bulging case, she ate a breakfast of cereal with tinned pears. They reminded her of the gnarled tree she and Manny had planted in her childhood days.

'No, I can't ring my cousin. He would've left home by now. Never mind. Perhaps he'll arrive soon. If I leave my coat and things by the front door, I should have time to finish these notes of condolence and find my doctor's letter, before our postman returns with his home address. Malcolm won't mind delivering them to our friends. Manny would expect mothing less of me.' A tooting horn alerted her. 'It sounds as if my cousin's car has arrived.' Walking across the creaking floor boards caused Gigi to shudder. 'This rickety old house makes my blood curdle. It really is creepy. Strong winds rattling the windows sounds like a banshee wailing on a winter's night.' Collecting her case and things from the doorway she cursed the weather. 'I'm anxious to begin our journey down through to the Cotswells.' On entering the car an unexpected storm deluged the city, preceded by the blackest-convulsing clouds.

In Gloucester, Ian Ross impatiently tapped his shoe on the brass railing in the Dorchester lobby. Scanning his watch for the umpteenth time he stomped the same black brogue hard on the polished terrazzo floor.

'Where the hell could she possibly be?'

He glossed over an article in the local paper. Rechecking the time on his father's gold wristwatch it compared favourably with the hotel's wrought-iron-framed wall clock.

Unconversant with this town, Gigi's cousin grizzled over being delayed in endless streams of traffic. His ancient Ford idled behind vehicles banking up with every mile they travelled. Frustrated, he preferred not to proceed further into the city.

Depositing his passenger on the nearest corner to the hotel, where Ian had booked in, he set all her worldly chattels down on the wet pavement and waved goodbye. Gigi battled through crowds. Slowly she humped her things up several steps leading to the first landing. From there a porter relieved her of two heavy cases. Giving her a courteous nod, he carried them up the final three steps.

In her clutch bag were two keepsakes. A small porcelain figurine and an antique cameo-locket on a black-velvet choker, which contained a lock of saffron hair. Treasured legacies of Manny.

7

Prowling around the hotel foyer in a glum mood, Ian's sullen look frightened Gigi as she approached the desk.

'You know how I dislike wasting time. Explain your late arrival,' he snarled, sarcasm seethed from twisted lips and eyes belching with fire.

The harshness of his abuttal caused her to miss a beat. She found it difficult to answer civilly. 'Ian, I…I have missed you. Why didn't you keep your promise to ring me?' Fearing a backhander, she hesitated to embrace this man who resembled a fire-spewing dragon. Petrified by his glumness, she dropped the hatbox and make-up case. As her wafer-thin arm clasped the thick winter coat around her shivering body, she sobbed piteously.

The dragon reared its ugly head. Sprouting fire, he stood back and nudged her free arm away from his person.

'Must you create a scene? You're making a show of me.' His low mutter sounded louder than an explosion in her ear. 'Fetch your hand luggage and follow me to the lifts. I have something to tell you, once you've settled in *my* apartment.'

In his customary abrupt manner, he directed a hall porter to carry her heavy ports to his suite on the second floor. Reaching the door Ian deposited a lavish tip in the youth's hand in front of Gigi, who gasped at the princely sum.

'Haven't you missed me, even a little?' she whimpered, fearing to touch the coat sleeve, concealing strong muscles of his nearside arm.

'Hush,' he snarled, and gripping her fingers he gave them a firm twist. Watching the porter depositing her cases in the lift, his fierce scowl cautioned her not to say a word. 'How many times must I tell you to behave?' This illegible utter she alone could hear. 'Keep your mouth closed until we're alone. I won't tell you again.' A wrathful glower penetrated this frail woman's defensive shell, as she chomped on a broken fingernail.

On hearing the lift descend he spoke in a modest tone. 'I have a lot to tell you. I know my ideas will please you. Yes, I have missed you ... a little.'

'Sorry we arrived late. Mathew's car was delayed in traffic and by a series of unfortunate punctures. Changing the front tyres took ages and he wasn't at all pleased ...'

Interrupting, Ian scowled. 'I'm not particularly interested in your relatives. Nor do I appreciate marching around a second-rate hotel lobby until my feet ache. This place was all I could find on speck.' Holding one hand, he glanced down her pooling violet eyes.

She sniffed again. Although her timid smile lacked substance.

'My apartment has an exquisite view of the lake. Gigi, you'll adore being here with me. This suite of four rooms cost me a week's pay, far more than I was prepared to offer.'

Feeling alienated with a conglomeration of people bustling along the busy hallway leading to their suite, Gigi tried to conceal a pitiful whimper which caused Ian to relent. A brief peck on her flushed cheek and he unlocked the double doors. Rudely brushing her aside and without a word, he hurried down a small hall to their bedroom. Her heavy luggage had arrived.

'I wondered where my two ports, make-up bag and umbrella were. They must've come up in the first lift.' She skipped to the bathroom in urgent need of its facilities. Gigi now realised a new phase in her life was beginning. She reveled in the idea that Ian might be more attentive to her plight, once his self-indulgent mood mellowed. She began to undress and chose a watermelon woollen frock and clean underclothes to wear. A hot shower would wash away the grime of travelling long distances in fourteen hours.

Nestled on the sofa with his arms around her, Ian considered their time together was too precious to waste chattering. A sex-starved addict he needed his fix now. Not later, after lunch or tonight.

Rising to his feet, he swung the electronic arm over to rest on a seventy-eight disc on the turntable. Settling the needle on a record Ian returned to cuddle Gigi. The lilting melody he'd chosen lent an ambience of love to the dismal room. Crooning to the tune he smiled.

'Soft music creates the ambience for romance,' he whispered, 'let's dance.' They sedately swayed to the rhythm of a Bohemian waltz, until it concluded.

Anxious to fondle her naked breasts he demanded, 'Remove your clothes. I want to make salacious love to you now. You haven't the slightest clue how the strain of five weeks without you has made me feel depressed. Each hour of every day has dragged agonisingly slow, knowing you might be in another man's arms. Most nights I read to keep a languorous mind busy, not to brood in the loneliness of this ugly world without you.'

His fingers began to undo the buttons of her blouse with the idea of fondling her breasts. She pushed his hand away and gasped in fear of being sick.

'Ian, I'm nauseous and I haven't eaten all day. This squeamishness is awful.' She gripped her stomach to ease the cramps. 'I need to rest. After my shower we can talk. Darling, will you make me some tea? Weak and not hot will do… to take my tablets. The pain swirling around in my tummy is so severe, I can't breathe.' As she rushed to relieve its contents he gripped her nearside arm.

'You never listen to me. You *will* undress now. How do you expect a man to make love to you in stockings or damn corsets? Strip everything off. I won't warn you again.'

The pains were severe and her nausea increased. 'Ian, I need to visit the toilet. If I vomit or break wind here, it'll embarrass us both.'

She reached the bathroom in time. Exhausted, she struggled back to the bedroom and collapsed on a sofa.

'You'll be fine in a minute. That nauseous feeling will pass if you lie on the bed. I'll ring down for refreshments. Then I'll quench my thirst for you.'

'After vomiting for five minutes…you expect me to eat? Even the thought of food makes me feel worse.'

He refused to listen and ignored her plea to rest. A long yawn indicated how tired she felt. Insistent, he again ignored her and disrobed. His tie, shirt and trousers flew across the room and landed on a divan. 'Take off your blouse now. Or I'll remove it and your undergarments.'

In fear of another backhanded thump, she stripped down to her panties and brassier. Doubled over with stomach pains, she refused to argue with the intolerable grump.

'Gigi, must you taunt a man by trying to make him subservient to your whims?'

Indignantly she protested over being bullied. 'You're wrong, Ian. Please listen to me for a moment. I'm worn to a frazzle. And it's been a stressful day. You could try to understand how ill I am. I've been travelling longer than ten hours in a hot car and I need rest. I'll tell you my news and listen to yours, only after *you've* eaten. Then I'll let you love me. That's if your temper cools down. Why are you always so domineering and stubborn? And you're ready to pounce by humiliating me every time you *think* I'm guilty of some stupid misdemeanour.'

Naked, he faced the bedroom mirror and posing as an athletic model Ian admired how his pectoral muscles flexed and tightened.

Sauntering over to the massive bed, he pulled Gigi down on top of him. Exhausted from vomiting, she lacked the willpower to resist his sexual foreplay. She relented as his fingers began to fondle her naked breasts.

'Don't you love me? It's been five long weeks of abstinence and I need the love of a woman whom I can trust, now that I have the urge to copulate. What an unromantic notion to imagine you may've let another man caress you in *his* bed.'

Insulted by his uncouth remark she pushed him away. Resentment building within her overflowed and she glared at Ian, whose unnerving, insipid-blue eyes glowered with anger.

'I was going to tell you my news. It can wait until you apologise for saying such a despicable thing. I never have, and never will sleep with another man. I don't know where those weird ideas of yours originated, Ian.' Disgusted by his repugnant pout, she ignored him. A quick trot to the loo eased her pains and made her feel as though she could cope with more of his insulting comments.

'Before you started to ridicule me, I wanted to tell some wonderful news…'

He cut her short. 'If it's about your granddame's funeral, I'm not interested. My news is worthwhile. Yesterday I bought a house in the

Cotswold Hills. A limestone cottage with a thatched roof, surrounded by a magnificent garden. You'll adore *our* new home …'

She took the chance to interject.

'Why didn't you phone or write to me? Every day I expected a letter or something. By nightfall I realised you didn't care about me, or my health.' Doubling over in severe pain, her voice trailed to a whisper in his ear.

'You're not! No, you can't be? We're not married and you've a brat on the way. Why didn't you use some form of contraception? How far along are you?' Akin to an enraged bird of prey ready to devour a meal, Ian fumed and pushed her to the edge of the bed. Incensed by his sardonic glower, she threw a pillow at him. She disliked the savage twist of his downturned mouth which resembled a snarl and he scared Gigi.

With an indignant look, she pulled away from his reach and, burying her face in the pillow, she wailed pitifully. She could neither think, nor respond.

Ian gripped her arms and began to shake her violently. 'I asked how far you are,' he roared. 'You must be well over six weeks, if my calculations are correct.'

'The doctor in Llangollen recorded on my chart that I'm two months. I thought you'd be thrilled with becoming a father. I refuse to have an abortion, Ian. If you don't want our child, leave now. I'll manage to rear this babe in my womb, without you.'

'No, you won't! I'll take you to a hospital first thing in the morning.' Typical of a desensitised louse, he paced the parquetry floor incessantly.

Gigi flew at him. Enraged, her balled fists pounded his chest. 'You're not taking me anywhere. My doctor has given me a referral letter to a specialist in London.'

'Pull yourself together, Gigi. You're being stupid. Forget the London nonsense,' Ian scoffed as she tried to edge away from his clawed fingers. 'I can't leave work now. If you insist on keeping this child, you *will* give birth to it in the Cotswold's. Personally I …'

'Don't you say it, Ian. I'll hate you forever, if you do! I am going through with this pregnancy. I will never consider an abortion. Stay with me if you must, or leave. I don't care which.' Shaking with anger her voice trembled with defiance. In agony, she edged her bottom off the bed, tangled bedding broke her fall. He caught her before she hit the floor.

Until her eyelids began to flutter, Ian held his breath. Sighing, he withdrew his hand from under her neck. Relieved, his gaze focused on her misty eyes of gentle violet.

She spluttered, but couldn't speak. Something burnt her lips. A hottish liquid trickled down her throat. Forcefully, she pushed the glass away.

'What were you trying to do? Throttle me!'

'No. Of course not. You needn't panic, throw a tantrum or go into hysterics.'

'My throat hurts. Let me rest. What did you pour down my throat? I can't breathe.'

'A nip of whiskey. A little drop won't hurt you. It'll steady your nerves.' She gasped and found it hard to breathe. 'A woman should know the time of month when she's ovulating, or could fall pregnant. Why didn't you use a diaphragm? Surely your doctor suggested a condom as a precaution?' A plume of spit from his lips doused her eyes. 'Men don't understand a woman's menstrual cycle. Hell, what a mess you've placed me in now.' Cursing, he felt like throttling her. Instead, he clasped her hand to pacify a rapidly beating heart.

The sharp thump striking his kidney emphasised her anger. 'I suppose you'll tell the hospital doctor that *you* are not the *father* of *my* child.'

'Now there's a thought.' A hard kick in the shins made him gasp. 'I didn't mean the crude way it sounded. Gigi, it seems you take pleasure in shocking me. The last thing we need now is child without a marriage certificate. *Another lot of unruly kinder will ruin our marriage, she must have an abortion.* He scowled. A hard swallow had prevented this thought from escaping his lips.

'You make love to a virgin and expect her to understand sex. Because, that's all love means to you. Sex to satisfy your lustful desires. I've news for you Ian David Ross.' Both hands cradled the slight swell of her tummy. 'This *is* your child. Our baby came from your loins. You should love our boy, as I will do.' Having removed herself from his presence she called from the bathroom, 'If not, leave before I come back to the bedroom.'

Ian looked at his crumpled shirt and trousers on the divan, then at his leather belt. *I feel like beating some sense into the stupid bitch. She must know her menstrual cycle and the dangerous times. Every woman does. I'll allow her to continue with this damn pregnancy, on one condition. She will not deny me my rights as a husband. The fool, she's adamant not to abort the fetus and it's not yet drawn a breath. She thinks it'll be boy. How ludicrous.* His silent ramblings were ceaseless as he dressed.

Pensively he surveyed the view from their bedroom window with sun glinting on the northern mountains. Shadows on their high snow-capped peaks and sheer walls looked bleaker than he was feeling. He glanced up at the whispering fields of rye grass, a passive green sanctity of the high ridged plains.

'Well Ian, what's your decision? You either stand by me all through my pregnancy, or you know the alternative.'

'My love for you far outweighs a stupid argument. I have no intention of leaving you. I'll let you keep this child, provided you take care of it. I prefer not to touch or nurse it. I will, however, support you both financially. This my final word on the subject.' Her glare indicated she needed to say something. The disparaging look he gave her held a threatening connotation. 'Don't you interrupt me, let me finish. Tomorrow, I'll drive you to the hospital. You'll see a doctor, any doctor and book in to have the baby.' Donning his suitcoat and tie, he snarled, 'I don't want you whining that I haven't performed my fatherly duty. Time might mellow my attitude, if it means losing you.'

'I want to know why you weren't pleased with my news. I thought you would be ecstatic like most first-time fathers. Am I selfish in wanting to keep our child?'

'No. We shouldn't be contemplating a family until we're married and settled in *my* new home. The time we spend together is precious little. You must agree I am right, Gigi. Seldom, I'm wrong in the matters of the heart and you know it.'

By wheedling into her good graces, he was deviously planting a seed of doubt in her mind. *Why should I play second fiddle to an illegitimate brat? One I didn't plan or want. If I disclose a slither of impropriety she'll feel guilty about her neglect of me and my needs. A child will be a catastrophic climax to a dramatic, damnable life.*

'Gigi, it may not be the right time to continue with this pregnancy. A young mother, you could die giving birth to the child. No infant is worth your loss to me.'

'You don't want this babe. I know that by the way you keep hounding me to get rid of him. No one has the right to deprive me of holding my infant son in these arms for the first time. Don't ask me how I know our babe will be a boy. I just do.'

Bored with her prattling he didn't reply. Her answer came in the form of a smirk and in angry eyes. A minute later, Ian condescended to speak in a calm tone. 'Listen and don't interrupt me again. I refuse to discuss this matter, until you've seen the doctor.'

Hurt, she stormed off in a huff to fetch her galoshes, mackintosh and brolly from the utility room. Alone she wandered by the lake, letting a river run down her drawn cheeks. The tears fell on passive waters. Sniffing, she fed the swans with sedge-weed and watched the signets flapping their tiny wings as if ready to take flight. Wearily she sighed and craved Ian's company. *I will never forgive him, yet I can't be angry. He deserves my love and I was a bit nasty. If I let him think I'm wrong, or give him a hug he might relent and then forgive me. I regret our disagreement, it should never have happened.*

Enthralled by the carnelian-crested flamingoes settling on the water's edge, she sat on the grass. As they waded in deep water to fish with their curved bills a swarm of John Dories scattered in all direction. Amused by their antics, she rollicked with laughter until tears of joy ebbed from her swollen eyes. Leaving the birds to feed their offspring, she closed her brolly then skipped over the wet cobblestones and giggling, she trod in puddles, until water sprayed everywhere.

On her return to the hotel an hour later, she searched every room. Her bird had flown. A hot shower eased her stiff shoulder muscles and aching legs. Feeling refreshed in spirit, a rest invigorated her jaded nerves and settled a squeamish stomach.

5.pm: Clutching her door key and brolly, she left a note on the reception desk. A passive smile embroidered her unadorned lips and with outstretched arms she embraced the cool, magnolia-scented night air. She stood in a bazaar to sniff sensual spices from the orient. Bought a lime ice-cream from a vendor and watched Oriental ladies modelling for a film crew, in their exquisitely silk-embroidered kimonos. She admired their bouquets of tuberoses, lemon-tinted carnations and mock orange blossoms. Hearing the chimes of a church-tower strike six and feeling tired, she sauntered back to the hotel.

Two hours earlier: Putting on his raincoat, Ian prowled the bedroom in endless pursuit of an answer to their predicament. Sitting on his haunches, he held an empty whiskey flask. Distracted by the blast of a fire siren he listened. *She hasn't returned from her walk. Why did she have to be so dominant when she threatened to leave me? She will abort the child. I'll make sure it will not live to draw its first breath. An accidental fall down the lift stairs will do the trick. It worked before with my bitch of a wife. I'll bide my time until we're alone. No simpleton can compete with a mind as powerful as mine. Blast it and her. I'm off to the hotel bar to imbibe on matured malt whiskey. She can go home to Wales and stay there. I don't care a damn what the bitch does.*

They passed in the hall. Gigi exited the lift as Ian walked to the top stair. Alone in their rooms she sighed.

'Manny's doctor in Llangollen assured me there would be no risk, as long as I rest every day. I'll not let him bully me into having an abortion. Ian acts as if he's the son of an imperial potentate when he's riled. I'm strong-willed if confronted with danger. Has he concocted an evil scheme to make me surrender to his dominance? From today on, I'll keep a few paces behind him, so he can't bully or hurt me.'

Yawning, her thoughts digressed to the day of Manny's death.

'Doctor Howlett gave me his blessing and warned me that I could miscarry. If there's a show of blood, I must seek help immediately. I'll catch the first bus to London in the morning. Ian will probably growl or abuse me, if I tell him of my plans.'

Edging herself further up the bed, she switched off the lamp. *Without him nagging me, I should get a full night's sleep. God, please keep him safe, and away from me. I hope he's sober when he comes in at some unearthly hour. If I'm asleep, he better not disturb me.* Logic had returned and with less stress, her sedated mind settled to dream.

Numb with indecision, Ian hadn't anticipated her previous outburst of temper. *What will she say if I roll in drunk?* For once Ian Ross was at a loss for words. *Eventually she will relent and learn the meaning of obedience. If sex is what I desire, the stars above will be my judge, not God. I'll have my way with her before dawn, with or without her consent.*

Swaying on unsteady legs he heaved on stepping in the lift. *A drink will refresh a dry palate. Fortified by a good nip of malt whiskey, I'll be able tolerate the bitchiness of her temper.* This boosted his courage and egotistical image. In a despondent mood Ian had ignored the sanatorium doctors' caution not to imbibe. Should he stumble on the path of non-indulgence again it could mean his death. Nevertheless, a proverbial drunkard he ignored their warnings.

Stepping from the lift on their floor his slobbering lips puckered.

'If she denies my love, I'll commiserate in the bar.' Ungainly swaying, he upheld the bottle of Scotch. 'This might tide me over until I'm in that bitch's good graces.' Concern for her welfare became distorted on slinging a couple of gulps down his throat. A rambling tongue couldn't be hobbled. 'Who does Gigi, a tramp from the backwater of Wales, think she is by blaming me for everything? If I can seduce her, any man can. That mongrel Pierre will only make love to her over my dead body.' With each swig of whiskey, his belligerency coupled with a throbbing head increased. Almost legless, Ian managed to stagger along the hall to their suite as a church bell chimed midnight. His wallet, now greatly depleted, fell from his hip pocket. The barman had tucked Ian's twenty-pound tip in his overtime jar.

Day was in the throes of breaking when Gigi awoke. Prevented from moving, she lifted a naked arm from her breasts. The overpowering stench of whiskey mingled with vomit expelled from a gaping mouth caused her to heave. She reached the bathroom in time.

'I feel sorry for Ian. I may've been a bit nasty breaking my news to him. If we and our babe are living in harmony in the Cotswold's there shouldn't be the temptation for him to drink heavily. How devastated he looks lying there. His pallor is enough to frighten a ghostly apparition. Not since the night before our departure from Highgate have I seen him looking so pathetic. I wish he'd awaken. I need to tell him that I still love him.'

Her meanderings ceased on him stirring, although still in the land of slumber he didn't awaken. *The stuffiness of this room is suffocating me. Yuk! The smell of his breath is vile.* Feeling squeamish she darted to the balcony. The cool air cleared her lungs and settled a grumbling stomach.

Neither storm nor tempest could dampen her spirits more than Ian had done the previous night. Unpacking her things, Gigi put on a wool-lined coat and struggled to fit muddy galoshes over her shoes. She searched for the latch-key. This proved fruitless until she spied a set protruding from his trousers. Fearful of disturbing Ian, she crept from the room. With his keys tucked in her raincoat pocket and with a thumping heart she hurried down to the lifts.

Alone on the hotel's steps she took a couple of deep breaths. 'I couldn't cope with an argument this morning. Rest is what he needs. The longer he sleeps the less chance of a flying elbow collecting me in ribs. I don't think he'll remember thumping me last night. My lips are quite swollen and sore. He must've been dreaming of being in a fistfight with someone.' Her greatest fear lingered in a distraught mind. *If he hits me again I could miscarry*. She recalled having twinges of pain in her abdomen during the dark hours, when sleep had eluded her.

Now as she clutched her stomach, Gigi surmised it might be the "quickening". As a teenager she'd often heard Manny use this medical term.

'Morning sickness is the blight of my life. My doctor told me nausea could be my constant companion until my sixth month. It sounded absurd then. Now, I'm certain he spoke the truth. This horrible sickness lasts all day. If Ian could feel the weird sensation rippling through my body, he never again would be critical of my delicate condition.'

No one was within call when Gigi ceased criticising Ian. Cautiously she moved down the hotel steps.

'A brisk walk will ease this melancholic feeling,' she sighed, taking in the cool air. She loved strolling in misty weather with birds on the wing cooing their lullabies to luckless winds. In silence, she repeated the walk of yesterday as a plover swooped into the water. Without warning, it resurfaced and soared high in the cloudless sky. Heavy rain clouds had expelled their load and moved on. The mist couldn't dampen her spritely steps, and her miseries faded into obscurity.

Earth's dark canvas by morning light, reflected a myriad of greens as trees caressed the motionless lake. White-breasted egrets caused ripples on its surface. Meditating, she counted the fledglings as they glided past their parents. Birdcalls were on Nature's agenda and Gigi anxiously awaited the park's wildlife stirring.

In a reflective mood, she surveyed driftwood and waterweeds as they swayed on the ebb tide. Dying autumn leaves glowed in sunlight as a zephyr swirled them to greet the sedges, where they danced to a mythical tune. As the wind increased, gold to rustic calendula petals fell, sending their tattered emblems soaring high above her head. Swans fluting their wings reminded her of graceful sea anemones swaying on a millpond of dreams.

Strolling up the grass she supported her tummy with gloved hands and wondered whom her babe would take after.

'A baby boy with dark curls and who will grow tall with masculine features is my choice.' His eyes, she hoped would be violet, not like his father's insipid eyes. She found it hard to tell if they were blue, or a sombre grey. At times they were expressionless and deflected the colour of his clothes.

Resting to catch a breath on the hotel's stone wall she reminisced. *Last night when I tried to awaken Ian, I saw a hateful look in his eyes. And his facial contortions were grotesque. He scowled and mumbled something about the baby not being his. I moved back, because his breath stank of garlic. I often sense that he hates me, especially if he's drunk and thinks I've been unfaithful. He is very cruel. And his ugly threats sounded like German, or some kind of dialect. Ian embarrassed me in front of the housemaids, who came to clean our rooms, by calling me unmentionable names, words I'd not heard before. He raised his hand and I thought he would strike my face again. The black bruises and swelling under my eyes will fade in a day or so.*

Her mind ceased meandering, when a mother and child sat on the wall beside her. Gigi smiled at the five-year-old girl who offered her a sweet.

'No thank you darling. My lunch is ready.' Excusing herself, she walked up the hotel steps and to what she imagined would be a day of dejection or misery.

Stepping from the lift, Gigi tripped on something and fell. Stunned, she looked up at a smirking face. 'Why are you lying there? Here, let me assist you to stand, my dear. You ought to be more careful, or you could abort the child.'

'You tripped me on purpose, Ian. Why?'

'No, I'd only just walked up the stairs when I saw you fall. Why do you always blame me, when you know I'm innocent?' He raised his arms. 'There you go imagining things again. It never crossed my mind to hurt you. Grow up and act your age. Follow me along this corridor and stop whining. I've had enough of it and your stupidity.'

Gigi refused his offer of help. Climbing to her feet, she walked past him. In their bedroom she threw her coat, scarf and gloves on the bed. With the sudden urge to pee, she rushed to the bathroom. She rinsed her bruised hands and knees, discarded her torn stockings then lay on the bed. She dreaded facing Ian. 'His crude remarks and a surely scowl frightened me,' she wailed, amid tears. 'I hope he will let me rest a while. Ian may, in time, see how his horrid tempers and cruelness are affecting me and the babe.'

8

A week after her arrival in Gloucester, she and Ian moved into their picturesque cottage at Evesham in the Cotswold Hills. The home was comfortable, yet sparsely furnished. Its peaceful ambience increased her enthusiasm to design curtains and bedspreads, allowing her to pursue her artistic talents without hinderance, and it enhanced her dismal moods.

Within a month, she and Ian had tied their nuptial knot in Stratford's registry office. Their honeymoon night they spent in the Hound and Crown Tavern, a short distance from the small stone cottage where William Shakespeare was born in 1564.

Early the following morning, they began their return journey to Evesham. A sudden downpour of rain couldn't dampen his enthusiasm to be alone with his wife, in their new home. Conflicts over their babe were put to rest. Ian had relented and no longer insisted on her having the babe aborted.

She doubted her husband's motives. Sullenness distorted his handsome features, especially if the topic of her pregnancy arose.

Gigi carried her baby through to the eighth month. At five on a freezing mid-winter morning, a neighbour drove her to hospital. On the thirteenth of December in 1947, she gave birth to a boy. Ecstatic over her son's arrival, her face showed intense anguish when the doctor enquired if the infant's father might phone or call in person.

'My husband works long hours teaching pupils at a public school in Stratford-on-Avon. He's a very busy man, although he may drop by tomorrow. Not tonight. Most evenings he lectures to students in the art of sketching wildlife. Doctor, I never meddle in his plans, or ask what his agenda entails.'

Over a period of days before the birth, Gigi had jotted down what to say, if for some unknown reason Ian wasn't available. She wanted him to be there for the birth of their first child. After breastfeeding her son, she watched as a nurse wheeled his crib to the nursery with tears welling in her eyes. In bed, this young and inexperienced mother scribbled a note to her disorganised husband and then cried herself to sleep.

Her babe weighed seven pounds at birth and measured twenty inches. He survived, in spite of his father having pushed his mother down the laundry steps a week earlier. An argument had occurred due to discord in the marriage. In a drunken stupor, Ian had lashed out with his fist and hit Gigi in the stomach. Paralytic, he collapsed on the laundry floor.

'From the moment of his conception, I named my dark curly-haired son Kendall, after a young dancer whom I worked with in Paris. Although sadness has blotted my life, I smiled when Kendall first suckled on my breast. How proud I'll be if he grows to manhood with the serene nature of his namesake.'

The nurse smiled. 'Kendall's a manly name. It suits this babe in my arms.'

'My prayer is for him to be healthy and possess my friend's loving nature, pleasant disposition, spontaneous wit and his enchanting character.'

Gigi unbuttoned her nightie, as the nurse placed Kendall in her arms and his tiny lips began suckling on the left nipple.

'Your son is a delightful babe, he seldom cries. You should be very proud of him, Mrs Ross. When you've finished feeding Kendall, I'll show you how to change his nappy. Then we will wheel his crib back to the nursery. The nursery window will be open from six for an hour, for your husband to see his son.'

In the past Ian Ross didn't given a tinker's damn what name to call him. He called him the brat, as he'd referred to Kristian, his firstborn child in Hamburg, Germany.

Gigi fretted for her husband, who'd left home three days before she was due to give birth to Kendall. A note left on the kitchen bench read: "I've been offered a teaching position in Stratford-on-Avon. This chance of a lifetime I can't afford to miss. Be home in a month, I". There was no endearing or carefully chosen words to indicate love. The single initial infuriated Gigi. A neighbour finally located Ian Ross in a Stratford bar where he insisted on buying a magnum of whiskey to wet the baby's head. Any excuse to imbibe, to hold a glass and end up legless, all of which she feared and disapproved.

A week later with a precious bundle in her arms, Gigi returned home to Evesham. After bathing Kendall, she dressed him in a warm gown of blue velour and kissed his bare bottom. She pinned his nappy and settled the dozing infant in his cot. Wrapped in warm bunny-rug and a cot blanket she kissed his forehead.

Gigi then relaxed in front of the blazing fire and watched its blue-tipped flames lick the chimney-bricks. She gasped on seeing a photograph of Kendall turned face-down on the mantelpiece. *Ian's done that to spite me because I refused to cook his breakfast. I wondered why he leered at me, before he left for work in Strafford-on-Avon yesterday. Will I give him a piece of my mind when he comes home on Friday? He's probably staying in a whorehouse there. A deceitful man, he doesn't know I found a handkerchief with lipstick in his coat pocket. I resent him touching me when he's drunk.* She shuddered at the thought of her husband lying next to her in bed after sleeping with prostitutes, or "women-of-the-night", in a dingy hovel or brothel.

'A week at home and he'll be gone again,' she moaned while checking on Kendall. 'I won't be sorry. He growls at everything I do. Ian accused me yesterday of doing more for you than I do for him. What does he expect? I a*m* your mother. It's my responsibility to feed and care for you. I love taking care of you and your needs, my darling son. I dislike and mistrust your father immensely. He blames me every time you whimper. But you seldom cry, only if you're hungry.'

An adoring mother, she slept in Kendall's room when storms raged, to protect him from the roar of thunder and lightning. Like most children, he was frightened of the dark. Gigi consoled her infant son, nursing him until he fell asleep, snuggled in her arms.

9

In London for a business conference on wool, Kirsten Svensson visited the Ross family. By stopping off for two weeks in the Cotswold's before returning to Australia, it enabled her to see Gigi and her infant and also to discuss if she needed any financial help.

Ian's surliness and cruel nature was the main reason behind Kirsten's sudden visit. From Australia, it was impossible to envisage how she and Kendall may be coping. Limited conversations over the phone prevented her from discussing any subject at length with Gigi. The fragile woman insisted things couldn't be rosier between her and Ian. Yet Kirsten doubted if this were true. She personally needed convincing and couldn't believe the rubbish that Gigi had ardently sprouted over the telephone.

Kirsten hadn't received a reply to her last four letters and it inspired her to remain in England for an extended period. Fearing she may have suffered a relapse after a shocking bout of pneumonia, or perhaps Kendall was ill again perturbed her. *Whatever the cause, it's apparently taken a toll on Gigi's health. Well, I'll soon find out the truth behind her lies.* Kirsten collected her luggage and tipped the cabby for his courtesy.

Pierre Bouvier and his son, Raoul, were frantically trying to find time away from their all-consuming family business to travel across the Channel to assess the situation for themselves.

'Everyone here is worried. It's so unlike Gigi not to write every few weeks. None of my British acquaintances, nor her friends have received a word from Evesham,' Pierre answered Kirsten. 'I feel relieved now you are there, dear lady. Please keep me my family informed of her progress. Kendall's illness has really perturbed me. Mercedes, Raoul and I send our love to you and Gigi.' The line from Montmartre, a suburb of Paris, dropped out. Stressed, Pierre let the receiver fall from his fingers and landed on its cradle.

Unable to pay their telephone account is a constant worry, Kirsten frowned. *She needs some means of communication in case of an emergency. Why hasn't Ian remedied this? I suppose he's too damn concerned with his extramarital affairs to care about his wife and child. He hasn't fooled me or Bjorn. And he agrees with me. I clearly recall Gigi disclosing his illicit love affairs and sexual exploits in a previous letter. That's when I discovered how unhappy she is in her marriage to that arrogant self-centred prig.*

What worried the Svenssons was how Ian Ross had threatened his wife in front of Mercedes. Two months previously Pierre's daughter had visited the family to see how Gigi was coping after being diagnosed with a serious nervous condition.

Kendall's godmother, Mercedes used the visit as an excuse to give him a belated birthday and Christmas present. She promised Kendall the gift long before his parent's phone was disconnected. Her short visit had caused a terrific row between Gigi and Ian. Distressed over the incident, she returned to France the same day. Before leaving, Mercedes overheard Ian forbidding his ill wife to have further contact with their family.

Loneliness was the reason for Gigi's nervous collapse. Neglect and a meager diet were a contributing factor of her health problems. Days would pass without cooking for herself or Kendall. With Ian constantly travelling on business trips, she grew complacent about everything. Slovenly in dress, she neglected to wash the parchment-like skin of her face. Long, matted and unkempt hair fell on her emaciated shoulders.

Ian refused to tell his wife where he would be at any given time. Weeks would pass without him appearing. Eventually on his arrival home he revealed, in a condescending tone, that he no longer loved Gigi. His condemnation of her was pitiful. Their latest argument had occurred over some insignificant triviality.

Kendall's morose moods increased every time his father came home. His irrational fits of brooding frightened Gigi who considered leaving home. Where could she go with a sick child? Fretting over not being able to confide in her grandmother, there seemed little hope on the horizon for either her or Kendall to lead a normal life.

Kirsten dropped by early the following Monday morning to pay her respects to the Ross family. She surmised with Ian away, it would be safe enough to stay for a couple of hours in Evesham. Her knock and calls were unanswered. She prowled around the house exterior checking on signs of life. There was none. Kirsten then turned her attention to the orchard behind a disgustingly overgrown garden. Everywhere she searched there was no sign of the property ever being inhabited. Wilderness had consumed the once beautiful garden. Nothing exuded life. Even weeds seemed out of place in this ugly expanse of untamed wildness.

Finding her pen in the remnants of a once tidy purse, she hastily scribbled a note then pushed it under one corner of their front door mat. There it couldn't be missed.

By removing both stilt-heeled grey shoes it enabled Kirsten to jump high enough to see through the partly open kitchen window. It proved a futile exercise. Floral sage-green curtains fluttering in the breeze blocked her view. She rechecked the house, and found its front and back doors locked. *Where could Gigi and Kendall be? This is weird.* Not one person lingered in the street for her to enquire of their whereabouts.

In this brilliant cloud-free day and with time to spare, Kirsten wandered down by the River Avon. A huge tree blocked her path. Sitting in the shade, she took a biscuit from her handbag. 'I'd love a cup of hot tea to quench my thirst. At home, the first thing we do if friends drop in for a chat is put the kettle on to boil.'

This courtesy was afforded to guests in most Australian homes, particularly in outback or country homesteads. In the land of wild kangaroos, wallabies, koalas, enticing beaches and rugged mountain ranges a pot of hot tea, or perhaps a cold drink would be made the instant guests arrived.

Australia was one main reason why she had flown halfway around the world. Kirsten and her Norwegian-born husband Bjorn, both naturalised Aussies, knew if given the option to leave England's bleak winters for Australia's sunshine, Gigi would jump at the chance. Ian would be the only resistant bug in their plans.

Watching a flock of geese on the wing Kirsty reflected: *We expected 'Lord Muck' to rebel against our offer to sponsor his family. A bane in everyone's life, Ian will probably object just to be nasty. Especially if he thinks our offer might please his wife. Kendall, whom he barely tolerates and openly declares his dislike for, is so young. How can a child of not yet one, understand why his parents are constantly bickering?*

Kirsten's mind travelled on a different avenue of thought. *Bjorn thinks Ian's a selfish and inconsiderate bastard. My opinion of him differs slightly. With Kendall and his mother under our care in Australia, we can monitor the family situation. I know Bjorn will make Ian see reason for striking his wife. I knew Gigi would be making the biggest mistake of her life by marrying that beastly man. She seldom listens to me or Pierre, unless it will benefit Kendall in some way.*

Comfortable on the riverbank, she snapped a reed of grass and chewed on its pulpy stem. A lesson she learned as a child in the bush, to ease one's thirst. *I must get in touch with her somehow, if only to relieve my own mind.* For an hour Kirsten mulled over the Ross family's dilemma, until she realised it was a waste of effort and time.

'Seldom I leave the confines of our local village. The milkman brings my groceries, blocks of ice and meat orders twice a week. I visit my specialist in Gloucester once a month', Kirsten recalled Gigi having written in her last letter.

3pm: She ambled back to their place to find her note undisturbed. 'Ian hasn't come home and discovered it,' she sighed. This risk she initially had taken in good faith. She wasn't going to tempt fate twice in one day. Her taxi arrived at the specified time for her return trip to Oxford. Destroying the note in the terminal, she discovered her London flight had been rescheduled to five pm.

The only consolation of her visit to Evesham was not confronting Ian, until his wife could contact him re their offer to relocate to Australia. Kirsten knew the Department of Foreign Affairs officer sanction visas, before allowing the family to emigrate. Hopefully she had laid down enough groundwork for Gigi to catch her husband in a receptive mood. Whether he would accept their offer of sponsorship was non-negotiable.

The second she stepped in her London hotel, Kirsten placed a call through to Paris. The prospect of speaking to Pierre seemed unlikely. Instead of driving herself to despair, she decided to read a travel brochure in a restaurant, adjacent to the reception area. The phone booths were located in this area and she couldn't miss hearing her name called over the intercom.

'Mrs Svensson, would you please return to the reception desk.'

'Ah,' she gasped, 'it's not before time.'

'A moment please, Madam?' the receptionist nodded. 'Thank you. Put it through.' She paused and placed several fingers over the receiver. 'The exchange is having some difficulty. The number you require is busy. Your call is being connected now Madam. Please return to bay one.'

Eagerly she spoke in the receiver. 'Mercedes, is your father home? Darling, this is extremely important, I must speak to Pierre. Yes dear, it is urgent.' The intermittent static forced her to wait. The line cleared. 'He's not, well you'll do.' She listened to Mercedes and replied. 'Thank heavens they're both safe. I've been frantic with worry over their disappearance from home. I'll be on the first available flight from London. Mercedes, don't tell Gigi that I've spoken to you. I'll give you a bell from Orly Airport, after I pass through Customs. Bye sweetie.'

Before replacing the phone receiver, she heard Kendall speaking to his mother. Mercedes, or her brother would meet her at the airport. Kirsten expelled a relieved gasp. Leaving the phone booth, her step quickened to reach the departure gate in time.

Within fifteen minutes she relaxed on the flight from Heathrow to Paris. Tired both in body and mind her flight taxied along at Orly runway on schedule. Cleared through Customs she hurried to where the phones were located. Suddenly, a firm hand gripped her shoulder, and a familiar face appeared in front of her. Blocking her path, Raoul relieved Kirsten of her hand luggage.

'How's my favourite *tantine* on this glorious evening?' Raoul embraced his godmother. 'Going by the flight number you gave my sister, this should be your luggage coming through on the carousel. You must be exhausted, Kirsty. Reaching here in near record time, you must've flown over on the breeze.'

'Raoul, it's terrific to see you again. Ha, I would've sprouted wings and flown over under my own steam to arrive on time. Actually, I was fortunate to get a cancellation in London.' Kirsty's faux-fur dropped on an airport seat. Quickly Raoul retrieved the cape and slung it over his free arm as she continued, 'I hope you've managed to convince Gigi to stay with you for another week. You didn't say a word…'

Raoul smiled. 'Our secret, I believe you mentioned to Mercedes. No, Gigi doesn't have the slightest idea that you're on the continent.' He collected her case and overnight bag from the carousel and together they walked to his car. 'Mercedes and I fobbed her off; by telling her we needed milk. We are good liars, and Gigi believed mine without question.' On reaching the Citron, Raoul held open the passenger door for Kirsty and put her luggage on its rear seat, humming a tune he returned to his own seat.

'How long were you waiting for my flight to touch down? Probably hours, if I know you. Gigi must think you're milking the cow.' Listless, Kirsty didn't smile.

Raoul laughed. 'I left home the instant you phoned the second time. I stopped in town to buy dad a new bowtie, on the way to Orly.'

'Thanks awfully for picking me up.'

'I couldn't think of a prettier pick-up than you, Auntz dear.'

Kirsty's shoe collided with his ankle. 'That's for being smart. And don't call me Auntz. That's what Kendall named me. Not you, now you're an adult.'

Approaching the Avenue des Champs Elysées she looked across at her French chauffeur. 'Tell me darling, how did she come to be in Paris? I was under the impression that Ian wouldn't let Gigi out of his sight. Raoul, he's a monstrously jealous individual. A man whom I could never take into my confidence.'

'My father went to London on business He couldn't contact Gigi by phone and wondered why. We're all anxious to hear about her and Kendall's health…'

Kirsty interjected, 'Talking of health. How was your mother when you saw her last week? Your father told Bjorn that Marion's in hospital. I didn't mean to interrupt you Raoul, but I am anxious to see the doctor's report.'

'My mother is now in a home for the dying. We will visit her tomorrow. Kendall's skin felt like parchment to touch, Dad said after he visited Evesham. Gigi looked ghastly. There were dark circles under her eyes. Bones were protruding through the skin on her shoulders. Pierre paid their telephone account. In no uncertain terms, he stipulated how worried we are regarding her health. Gigi refused to listen to him. That afternoon my father packed their ports, cuddled Kendall and kissed her. Gigi insisted that she wouldn't leave home or Ian. Dad knew she couldn't cope with an ill child and her emaciated body resembled a skeleton.'

The traffic was dense and Raoul cursed. Weaving through two and four wheeled vehicles, he eased his foot off the accelerator until the car stopped. Once the traffic began to move, or divert in different directions he relaxed. 'Ian rolled in right on dinner and had a terrible argument with Pierre. Inebriated, Ian said everything was fine. Drunk as a skunk down a rabbit burrow, he apparently didn't know to whom he was speaking. My father left there that night. The following afternoon, before Ian returned from Stratford, dad drove back to collect Gigi and Kendall with two extra tickets to Paris. I met their flight at Orly, and drove them home here to Montmartre.'

'Raoul, you can't imagine how worried we've been over their health. When the opportunity arose, I got a standby flight from Sydney to London. And you know of my misadventures in Evesham.'

'*Oui*, I do. Gigi's health is improving, because my sister cooks nourishing meals. I can't fry water without it burning. She and Kendall has been here for two weeks and we enjoy every minute of their company. We think it miraculous that she's still alive. It will take a long time for her lungs to recover after such a severe bout of pneumonia.' A roadblock up ahead forced Raoul's car to stop. 'Kendall is a delightful child. He cringes if anyone mentions his father. Don't say a thing in front of him about Ian. He'll run and hide if you do, Kirsten. It will take a while for you to understand his childish gibberish.'

'Thanks for the warning. I doubt if I'll discuss his father, other than in a business capacity.' Kirsty's mind steered away from what Raoul was saying. Realising this, she embarked on a different topic, one of mutual interest. 'Has your father mentioned to you and Mercedes regarding our offer to relocate their family?'

'Bjorn told him yesterday, over the phone. Dad said that you are willing to discuss their marital problems with Gigi. She may not confide in you or Pierre, until something definite is decided. Personally, I prefer not to be implicated in their mess.' Raoul swore in French. A near accident caused a delay at the next intersection. When able to concentrate he continued. 'Gigi is awfully unhappy. She won't admit it of course. One look at her emaciated features will put you in the picture. Kirsty, if you want my honest opinion, I think she's a fool not to divorce her husband. Ian sounds like a horrible person. Is he a Scottish-born peasant?'

'Raoul, I agree with you. Yes, he reckons he was born in Glasgow. He can't do any wrong in her eyes. That's why she refuses to leave him. I don't mind admitting, like your father, Bjorn and I are worried that Ian will harm her or Kendall seriously. It sounds as though the boy deeply fears and mistrusts his father.'

'Murder them both, do you mean? Kirsten, that's stretching the truth a bit. I think he is capable of threatening Gigi. Go any further I doubt it.'

Raoul thought it inconceivable how any man could murder his wife and son without flinching. He knew Kirsten and Pierre's opinion were identical. Raoul knew neither would fantasise about murder, or deathly threats. Yet he imagined the truth was more than a possibility. Gigi had led everyone to believe that she idolised her husband.

On the huge roundabout near the Arch de Triumph, Raoul sighed: *Dad's version of Ian's evil misplaced judgement differs from mine. Evil breeds fear and destroys a person's mind. In her pathetic state, illness has increased her fear of him. Ian must be a bounder to controlling Gigi with ugly threats. Our two intellectual minds think on parallel tangents. Kirsty and I both have psychic powers. We have sensed, over time, disastrous events long before they became a reality.*

As the car slowed at a smaller roundabout Kirsten tapped into his thoughts. 'Raoul dear, Bjorn and I know Ian *is* capable of murder. A couple of times we heard him abusing Gigi in our flatette. Bjorn was furious. He threatened to deck Ian, unless he desisted. Gigi was paralytic with fear. I consoled her. She was still shaking an hour after Ian had gone to bed. I thought he'd kill her that night. Next morning Bjorn found an empty whiskey bottle under the lounge Ian was sleeping on. My husband detests Ian. I won't say what he called him. They're words no lady would ever think of repeating in public.'

Raoul smiled. *I have often thought of vile words to call him.* Silent, this French musician continued to watch the traffic. *I know what I would like to do to the mongrel.*

'I've devised a plan to assist the family and Bjorn agrees. Gigi might give me a fair hearing. If she doesn't, I dread to think what will happen when she and Kendall return to the Cotswold's.'

Swerving in behind the block of apartments his father owned, Raoul pulled the Citron to a halt adjacent to its rear entrance. Before they alighted, he turned to face his godmother. 'Last night I overheard Gigi tell my sister that Ian is seldom home nowadays. It appears that he spends most of his time at Stratford-on-Avon. Teaching Gigi says, though

I doubt it. In my opinion his sexual liaisons with *la Fille de joie*, would be closer to the truth.'

'Raoul, I gather you, Mercedes and your father have the same opinion of Ian. I *am* aware of the reason why Pierre mistrusts Ian Ross. Most people detest him.'

'By what you just implied, it sounds as if you both dislike Ian. I think that miserable creature delights in being beastly to his wife. Gigi deserves to be treated like the lady she is, Kirsty. You must agree with me.'

'We all have our doubts. Who knows what goes on in his evil mind, Raoul? Ian is a womaniser. It seems he seduces women to satisfy his lustful desires. I hope Gigi hasn't suspected what we all assume is true. She will refuse our offer, if she even suspects him of being unfaithful.'

It is probable, Raoul surmised. His proverbial inner mechanism ticked as swift as a bomb ready to explode, where the Ross family were concerned.

She channeled into his thought again. 'What I intend to challenge him over will be a learning curve for all of us. Ian may not agree, knowing we've set in motion a strategy to get him to work long hours under a hot Aussie sun with little or no shade. Gigi can spend time either working in the homestead, or helping me with the bookwork. It will be a new field open to promotion. They'll adjust to keeping their noses to the grindstone in time.'

Kirsten removed her cream gloves, as Raoul spoke.

'Ian will find it hard working on a sheep station, if what you said is correct about the size of Jalna. Do you think he will eventually settle down and make it a profitable business? It will be the start of a different life for Gigi. With him being supervised by Bjorn and his men, he might stop drinking. If so, it should ease the burden on you and Bjorn. She needs to lead a normal life without the threat of being bashed. The deep purple bruises on her neck are fading and she often wears long-sleeved cardigans to hide the bruises on her arms.'

'It's sickening to think how cruel Ian can be when he's angry. Raoul, I'll try to talk her into staying on with your family. If that doesn't work, I'll take her and Kendall home to Evesham. Then I'll make sure they have enough food in the house to last a month or more. Ian might eventually see how his brutality is destroying their marriage and let her have a normal life in peace. Kendall needs a healthy mother to take care of him. Not an ill one, whose belligerence is capable of ruining his life.'

'Kirsten, you do know their phone was disconnected, before she came to Paris?

'Of course. I thought your father had paid the bill? If not, my first move will be to have it reconnected. I'll post a cheque to the company, so Ian can't argue with Gigi.'

'Dad has pleaded with her to remain here. Gigi can be stubborn at times. What I cannot fathom is why she wants to go home to Evesham and him. Dad has offered to let her and Kendall stay indefinitely in one of these apartments rent free...'

Kirsten went to open the car door and hesitated. 'Bjorn intends to keep a firm grip on Ian, most of the time. Neither of us will tolerate him mistreating Gigi. In time Ian *will* learn to consider her feelings. If he doesn't treat her with respect, Bjorn has threatened to warn him only once. You can imagine what the next step will be. Surely you remember your father's favourite quote. "Time is the redeemer of love, the builder of hope and of self-confidence." Kirsten paused to retrieve her handbag and coat from the seat behind Raoul. 'You know *my* pet saying if terror strikes.'

'Oui! Let peace reign and bygones be bygones. Dad often says that if he's stressed or angry. I wondered where it originated. Hush, I heard voices in the stairwell. It could be my sister and Gigi.' Nobody walked from the building. 'If the chance arises will you speak to her regarding his belligerent attitude? I'm not trying to influence you, Kirsten. I know you will make up your own mind, once you've seen Kendall and spoken to her.'

She gasped in horror on seeing Gigi for the first time in months. Leaving the building she didn't notice Raoul's car parked close to its entrance. As if on a transient plane, she turned and spoke to his sister. Mercedes swiftly diverted her attention away from the car and its passengers, by encouraging Gigi to walk fifty metres to the corner of their block of apartments.

Kirsten had crouched down in the front seat. Raoul looked in the opposite direction until his sister and Gigi were out of sight.

'If I walked past her on the street I wouldn't have recognised her, Raoul. Her body and arms are wafer-thin and her skin looks like parchment stretched over a skeleton. Gigi looks really ill.' Kirsten couldn't believe what she'd seen. 'Her pallor is greyish and she staggered along that path. We must get her away from this cold climate and soon. Or we may all be attending her funeral.'

Ready to step from the car, Kirsten hesitated to until they had turned the corner. Looking at her driver, several prospects flashed to mind. *I know two British diplomats who may be able to expedite their travel permits through in record time. Worth a try! Gigi looked ghastly. And I dread to think how her son looks.*

'If it's Kendall you're thinking of let him approach you. Remember to be tactful. The boy is edgy and becomes confused in front of strangers. He'll cleave to his mother's skirt, then you will never hear a peep from him.'

'Me be anything, but tactful?' she laughed. 'Raoul, I'll be polite in confronting Gigi and without losing my temper, unless she won't listen to me. I've learned to be patient in handling difficult people in my work. Ian has the beastly temper and an ugly trait of being insensitive, if he loses an argument. I've actually seen him carry out his threats. He and Bjorn have clashed over his drunkenness.'

Kirsty stood back until Raoul opened the door leading to the building's stairwell. 'Before I forget; thanks for billeting me. It saves me the hassle trying to find rent, or a flat in this beautiful though harsh city, noted for imposing strict rules on casual visitors.'

Nothing more was said about Ian Ross. But then he didn't rate a mention. It was obvious that his sexual exploits, had over long periods been conducted in Stratford-on-Avon, the English Bard's town. Here in this country, Parisienne bordellos were notorious for their discriminative acceptance of wealthy and foreign clients to accommodate ladies of all nationalities, from all quarters of this ancient city.

Unbeknown to the Bouvier family or Kirsten Svensson, Gigi had discovered her husband's philandering in Paris by accident. A card from a house of ill-dispute had fallen from his coat pocket. When she confronted Ian about his infidelity he emphatically denied the card had anything to do with him. Later, in a prudish mood, he boasted to Gigi that she lacked talent in bed like whores he wooed there. If the truth be known, he had described his sexual liaisons with indecent women in his son's hearing.

Kendall enquired in bed that night, 'Does Daddy love other ladies? Mummy, what's a whore? Doesn't he love you anymore?' How could she explain his father's sexual misbehaviour to a four-year-old? Her husband's sexual liaisons she loathed to believe. She couldn't ignore his illegal meanderings, especially after finding the card and also because they were constantly flaunted in her face.

Noon today: Kirsty suspected Mercedes had encouraged Gigi to go for a walk until she and Raoul were inside the building, knowing they wanted to surprise her.

Standing above the lift-well she heard Gigi say, as its ground floor doors closed. 'Your father must have come down from his penthouse, Mercedes. Or it could've been Raoul leaving for the Conservatorium. That's odd. I didn't see anyone leave the lift and only your family live in this building.' She then thought: *It could've been a tradesman, or the carpenters who are working in Pierre's observatory.*

Ready to fish in his pocket for the keys, Pierre acknowledged his son. 'I am eager to see if Mercedes and Gigi have returned from their walk. Raoul, hold these things please, while I unlock the penthouse doors. This lock has a habit of sticking. I'll ask one of the carpenters to oil it before he goes to lunch.'

'Mercedes uses her key. It works okay. I'm here to ask Gigi if she'll come on a drive with us. On such a glorious morning, she shouldn't stay indoors.'

'I doubt if she will go, Raoul. I bought *ma Cherie* these roses to brighten her day.' Pierre responded to his son in French. 'Please tell the ladies, I shall visit them around two this afternoon.'

'Dad, please don't come down to their apartment again today. You're tired after the long walk to the shops.' Raoul gave a second rap on the door, which his sister opened.

'Sorry, I was in the shower and didn't hear you knock. Gigi is lying down, come in both of you. If she asks where I am, Papa keep her busy. I'll be in my room wrapping a gift for Kendall. The kettle is on Raoul, will you pour four cups? You know how we all like our coffee. Fresh milk is the new refrigerator. Not the mini-one in the solarium.'

Gigi heard Mercedes speaking to a man and came to investigate. 'Where are you two?' Seeing them in the hall near her flatette, she bubbled forth with, 'Did you enjoy your walk, Pierre? I was lying down, as I'm a little weary after our walk around the block. I thought I heard Raoul's voice a moment ago?'

'My son, he is making hot drinks in the kitchen. You do look tired *ma Chere*. My dear, go back and rest in your room. Mercedes or I will bring you in your luncheon.'

'No thanks, darling,' she responded lethargically. 'You're a real sweetie for caring about me, Pierre.' Seeing the flowers her eyes sparkled with the gleam of diamonds. 'Are those apricot roses for me? Ooh, their petals feel like satin. They're beautiful Pierre.' She gave him a French kiss, a peck on both cheeks. 'I'll put these long-stemmed roses in water and then see what my son is doing. He can be a mischievous child at times.'

A timid little voice called from the hall doorway, 'I'm here, Mummy. They's in my bedroom, I founded them.'

Mercedes pushed past Kirsty and sweeping the boy up in her arms, she stuffed his tiny mouth with marshmallows.

'Shush darling. Aunt Kirsty is hiding from your mama.'

'Ooh, you can't come in Mummy,' he giggled. 'Aunt Mercie and me is busy.'

'Too late Kendall, I'm already her…e,' Gigi screamed with delight. 'Is it really you, Kirsten?' Brushing Mercedes aside, Gigi embraced her Australian friend. 'I can't believe you *are* here.' Afraid of the truth, she clutched her temples between both palms. 'Why didn't you let us know you were coming? Someone would've collected you from Orly Airport.'

'Don't I warrant a kiss from my kid-sister?' This endearing term they had used from their adolescence days, long before they began dancing together in Paris under Pierre Bouvier's guidance and in a professional capacity.

On the verge of tears, Gigi couldn't speak. Suspiciously, she edged back a fraction. Mystified, she glared at Pierre. 'There must be a good reason why you're here, Kirsten. Don't you start bullying me. I can't take much more worry. As of now, I won't answer anything you want to ask me. I wish you and everyone here will let me take care of my son and manage our lives without constant interference.'

'Hush for a moment and let me speak. Gigi, my flight to London was booked long before I posted my last letter, which you neglected to answer. Tell me why I shouldn't drop in to give Pierre and his family big hugs, after flying over two thousand miles across the Pacific? Finding you here in Montmartre is a bonus.'

Excusing his rudeness, Pierre interjected, 'Welcome to *Paris*, dear lady. I was only thinking of your family yesterday. How is your husband these days?'

'Pierre, let me say what's on my mind.' Kirsten butted in and apologised 'Then I can concentrate on telling you both my news.' This lie she knew would make Gigi listen.

Grunting, she remained aloof until Kirsten scowled. A caustic glower accompanied by a groan came though taut lips. Considering what she'd said, Gigi took a step towards her Australian friend. With tears ready to flow, she pleaded, 'Give me a hug, before you are cross with me again. Darling, have you been to my home in Evesham?'

'Yes, I did go there. When you weren't home, I assumed you were either here or in a hotel. On the way from the airport Raoul told me that you and Kendall were staying with his family. Stop sulking, Gigi. You look haggard and ill. I hate to think how Kendall looks. It's time you were honest with Pierre…'

'About what? If you're here to criticise everything I do or say forget it. You think you're an expert on marriage counselling. Well, you're not. Go back to Australia and stay there Kirsten. I can manage my own affairs without you spying on me. I'm fed up with people snooping into my life. I want to rear Kendall my way without everyone telling me I'm wrong.'

Suddenly it occurred to Gigi the real reason for Kirsty's visit. Puzzled, she tried to apologise. Instead, her mind reverted to their long association in the field of dancing with Sadler's Wells decades ago. *Why is she taunting me? I must be a fool not to have realised why she's here. Kirsten wants to dominate my life.*

'I know by the smug look that you and Pierre want me to stay here in Montmartre. Well, have I news for you. I *am* going to pack our ports, then Kendall and I are going home, where we belong. Tomorrow morning, I will ask Pierre to book us on the first flight from Orly to Heathrow.' Frantically waving her arms, Gigi pushed Kirsten aside and ran to her room.

Shaking his little head, Kendall looked at Mercedes with tears in his blue eyes. Confused, he wiped the tears on his jumper-sleeve. Whimpering, he watched his mother leave the room with a pen in her hand. *It's unlike her not to have told me why she travelled to England.*

Kirsten must have a good reason for flying halfway across the world. It's probably something to do with their sheep property in Australia.

Kendall followed his mother to her room. 'If I list what I need to pack, my darling boy, then we won't forget anything important.' He hugged her knees and smiling up at Gigi, his tears trickled down the collar of Pierre's navy-blue woollen robe.

A reasonable period of time passed before she returned to the sitting room to apologise to Kirsten.

'Now, we can talk about why you've decided to go home. Pierre's frantic with worry and appalled by your nastiness. He thinks he may have offended you, Gigi. And your hasty decision has something to do with your husband's brutality of you and this dear child. You are seriously ill. No, dying would be more precise. Those bones are ready to poke through the wafer-thin skin on your arm. It reminds me of parchment, not flesh. Your sunken eyes are bulging under hollow lids. Do you want to die and leave Kendall to fend for himself on the streets of Paris, or Evesham? Think of the consequences if he was abducted by a pedophile.'

'How can you say such wicked things? They're horrible, Kirsten. I chose to come here. Ian doesn't even suspect I'm in France. He thinks I'm in London with my cousin. Which you know, is untrue.' Her tone mellowed as Pierre's index finger touched his lips. 'It will be like old times bunking in together.'

'Well, it might be for you. Mercedes tells me we're to sleep in her room. Didn't you dear?' Kirsten winked at her Parisienne hostess.

This lie caught her unawares. 'Oh … oh *oui*,' Mercedes stammered, unsure what to say, or how to answer their Australian guest.

'I prefer to sleep with my son, thank you,' Gigi replied icily. 'Don't be offended Kirsten. The truth is Kendall won't sleep alone in a dark room.'

Excitedly, he clutched his mother's skirt and looked at her with tearful eyes. 'Yes, I will Mummy. Aunt Mercie can sleep in my room. Can't you Auntz? I'll be a good boy, if she can sleep with me.' Kendall's face beamed. He didn't know that he had placed both her and Kirsten in an awkward position. 'I told you they *was* here, Mummy.'

Cuddling the boy, Mercedes reminded him of their secret. She realised this situation was becoming far too complicated for his small mind to absorb. When she put him on the floor he shied away from

Kirsten. With a dubious frown and whimpering, he peeped from behind her skirt.

Raoul's warning rang true. Rather than force the child, she let Kendall approach. She knew it would be useless to discuss anything with his mother. Not in her rebellious mood. Expressing her frustration with a self-explanatory grunt and with a curt nod, she walked towards the bathroom. A cool shower would take the edge of her anger.

Refreshed and in a more congenial mood, she approached Pierre in the hall.

'Pierre, what I intend saying to Gigi, won't be for your delicate ears. I suggest you leave. It will be difficult for you not to interfere. It's imperative to make her understand how her temper tantrums are affecting you, and Kendall.' Never a good liar, she hated to deceive Pierre. 'This farce has gone on long enough.' A side nod indicated for *Mercedes* to take the boy to *her* room. Words and everything could fly when she verbally tackled his mother. Kirsten's demands were not for little ears to repeat, particularly if they were discussing his dominate parent. Their row must be resolved. If not, this confrontation could turn into an unimaginable argument of massive proportions. The outcome of which could be catastrophic for all concerned.

'Darling boy, I am not sure what to buy you for your birthday. You can choose a toy from the picture book in my room. Aunt Kirsty and Papa Pierre want to talk to your Mama. We don't want to listen to them talking.' Mercedes screwed up her nose and Kendall copied her antics, much to everyone's delight.

Pierre smiled as his daughter hugged the boy who chuckled. 'Mummy and Aunt Kirsty are noisy when they yell, Auntz.' Mercedes kissed his little fingers. And his smile broadened on his elfin-like face as she thought: *The childish name he's given me I cherish.*

Kirsty hesitated in the hall until everyone was out of earshot then she burst into Gigi's room. Straight to the point as always, she intended to tackle her head on.

Instead, Gigi got in first, even after receiving a glower from her intruder. 'I enjoyed our morning tea together. It will be terrific to hear…'

'Listen to me. I've not travelled all this way to chat over old times. Far from it! It's your health and that child's future we care about. My dear, you *will* hear my proposal and think it over. If not, our friendship ceases as of this instant. What is your decision?'

'Huh, so you do intend to be nasty. No Kirsten, I won't listen to you. Not until you tell me the truth why you've travelled thousands of miles to ridicule me.'

Bluntly, she would explain the motive behind her visit to Paris. Instead she paused to think: *Truth is the essence now. Fantasy has no room in either of our lives. Gigi will listen to my proposal, even to the point of being threatened.*

Comfortable leaning over the chair, she broached the subject previously discussed. 'Now Kendall can't hear us, I want you to listen to me. I'm willing to pay for your fares home and I'll come with you both, Gigi. You really have distressed Pierre. He and I are worried. If you say no and disappoint him I'll kick your backside…'

'I don't understand what you lot expect of me. My place is with my family. I won't divorce or leave Ian. I took my marriage vows seriously. Not on a whim, like you seem to think.'

Accusations began flying! She wouldn't listen to any proposition put forward by Kirsten. In a rage, she bellowed, 'I don't need you telling me what to do. This is my life. I have a child to rear without you or anyone interfering. My marriage to Ian is strong.' Her emotions bubbled over to a voluminous bout of raucous screams, accompanied by wild accusations. None of which made sense.

Kirsten poked her head into the corridor and nodded to Pierre. In a flurry of temper Gigi's foot collided with the door. It thundered shut. As it rebounded, her foot blocked their entry. She kicked the door back to prevent it from closing in his face.

'*Ma Cherie*, do not argue with Kirsten. What she proposes is sensible. Listen to her please. We are worried and want to help you.' Pierre pleaded with his ex-protégée: 'Her proposal will benefit your entire family. It sounds a profitable venture, I assure you, Gigi.' Anxious, his husky voice failed to escape through parched lips. Deeply concerned for her wellbeing and health, Pierre knew her rudeness was an effort to conceal a mass of underlying problems. Which neither relieved his anguish nor the stressful situation. Mopping his brow with a damp handkerchief he shook his head in disbelief.

Kirsty turned to face him. 'Let me handle this, Pierre. I have an idea how she might come to her senses and listen to me.' She patiently waited until his footsteps faded, then she re-approached Gigi's room and knocked on the door. Muttering incoherently, Gigi sank in a heap to the floor.

Helping her to disrobe, Kirsten assisted her to the bed. Pierre's folded dressing gown now lay over a chair. 'My girl you *will* listen to me, whether you want or not. Don't attempt to interrupt me. If you do, believe me, you will be the loser.'

Wide-eyed and mouth agape, Gigi knew she meant business. Sniffing, she gave her an impish grin and adjusted the nightgown on her bruised and aching shoulders.

'That's better. Now you're calm, listen to me. Bjorn and I intend sponsoring you and your family to live in Australia.' A finger warned Gigi to shush. 'It will take time to arrange. We have several options at our disposal to assist your sponsorship to take effect within a reasonable time. First, you must discuss our offer with Ian. You can reach me within a week. I'm taking you and Kendall back to London in the morning. At Heathrow I'll hire a car to take us to Evesham.' Up went her hand. 'No arguments. I shall instruct its chauffeur to stop at your local store. I'll purchase a month's supply of groceries, fresh fruit and vegetables, and a forequarter of lamb, pork, or beef. Once we've eaten a hot meal, I'll unpack your things, while you settle Kendall in bed.'

'Why are you doing this? My greengrocer delivers the milk, ice and bread every other day. I leave a note and money under the mat, near our front door. If I need meat he brings it the following day. Kirsty, I don't need your help. I *am* capable of caring for my child. Kendall never wants for a thing, not toys or books. Our doctor lives a short walk from home and drops my prescriptions in, on his way to the surgery. The chemist boy delivers our medications that day. Now you know we are a self-reliant family. Occasionally Ian brings home a meal from the local hotel or a delicatessen.'

'I'm not interested in Ian. He's the least of our worries. You are the problem. In time, I will teach you how to speak to Pierre without roaring or abusing him. The family has enough to cope with at present. Marion is dying, or have you forgotten? Consider their family for a change, instead of indulging and dwelling on your own miseries.'

She received a scoff instead of a reply. 'You seem to have everything worked out. How can I speak when you keep bellowing at me?' Her rudeness was ignored. 'Now leave and let me finish packing my son's clothes, and his birthday gifts.'

'Shut it, Gigi.' Kirsten snapped, on the cusp of exploding. 'I'll be in London for ten days, after I take you and Kendall home to Evesham. I do not want to speak, or see your husband until you advise me that he *is* in an affable mood. Then I'll talk to Ian.'

'Why do you and Bjorn want to help us? Neither of the men speak. They hate one another. You know that better than I do.'

'All that nonsense belongs in the past. You can't go on living in England. Look at your image in that damn mirror, woman! A scarecrow has more flesh on its bones than either you or Kendall. Neither Pierre or I want, or need to pay for your funerals. And that's exactly what we will be doing, unless you see reason.' About to be interrupted, Kirsty threw a shoe at Gigi, but purposely missed. 'You're ill, and I'm damn sure Kendall is too. The child is malnourished. And you're not much better. Admit it Gigi. At least do this for him, if not for yourself. Kendall needs you. What sort of a life would the boy have without you, his mother to care for him?' She drew breath, then finished delivering the lecture. 'We'll give Ian a well-paid job and train him in animal husbandry, and how to run a sheep station. It will be a profitable investment. You will all live with us on Jalna, in the manager's cottage. How's that sound for a fair dinkum offer?'

'You mean a genuine offer? Don't forget there is Kendall's schooling to think of. We can't afford to send him a private school.'

'Don't try to put obstacles in my path. You won't succeed. You should know by now how determined I am.' As her finger touched Gigi's chin, she didn't attempt to push it away. 'You can attend lectures with me on accountancy and mix with my social circle, men of wealthy standing in our community. Think over my offer while we're in Orly, or flying to London.'

Gigi glanced at Kirsten with a dismal look as an uncontrollable flow ran down her beautiful, though drawn features. A scarlet-flush glowed on her hollow cheeks. The bruises on her neck, abused by a drunken malingerer's hand, were less noticeable.

'I'm warning you Gigi, not to argue with Pierre. His patience is limited. Your insolence has become intolerable and he finds it hard to cope with your nonsense.' Kirsten's tone then softened. 'Surely you realise how your childish tantrums are tearing him apart. If you weren't so obstinate, you'd know how deeply Pierre cares for you.' Her angriness lessened. Gigi frowned over her suggestion. 'I didn't mean it sexually. He does love you, and like all of us, he is worried. Not many people I know will admit that. Think yourself fortunate that we do care for you and Kendall.'

'I don't know what to think anymore. Your proposal sounds logical, and profitable. It could end up being disastrous for my family. I'm confused. Can we reconcile our differences? If so, I will contact Ian. You know how stubborn he can be, if he's been drinking, Kirsty.'

'That dear boy,' her thumb pointed towards the room where Mercedes was minding Kendall, 'needs a better life than either you or Ian have given him until now. I'll kick your backside, if you spoil his future by not accepting our advice. Now sulk, because I've had a gutful of your pouting for one day.' Her aggressive tone conveyed how furious she was getting by her nonsensical platitudes. Kirsten considered it important and imperative to have stressed Kendall's sorry plight to his mother.

Kirsten walked to the door and clutching its porcelain knob she turned. 'You've always had good lungs, use them. Scream if you must, Gigi. Then perhaps you'll have regained some sense in that stubborn head of yours. You'll have plenty of time to think of our offer before I return to London.' Hesitating, she breathed in deeply. 'You forget I've worked my butt off among men in the field of raising wool. I don't intend putting up with your tantrums a second longer. Not when I won't tolerate nonsense from my staff or our stockmen.' Furiously she thundered from the room and pulled the door closed.

10

Before leaving his penthouse apartment to return home, Gigi confided in Pierre about her marital problems.

'Darling, I realise how stubborn I've been these last few days. Ian no longer loves me, I know that now. Kendall is frightened every time his father raises his voice or removes his belt. My son needs me to keep calm. I won't allow Ian to ruin our lives like he's done since the day of Kendall's birth. Ian tried to suffocate him with a pillow. I prevented it from occurring by striking him with my fists until they were bleeding. Ian laughed and pushed me against the bathroom wall. When I tried to stand he twisted my hair until I fainted. Kendall found me lying in a pool of blood. Ian must have kicked my face with his boot. A neighbour heard Kendall screaming. Neither my doctor nor the police could locate Ian. He didn't come home for two weeks. He won't get another chance to whip Kendall's little bottom with his belt. I promise you that Pierre. From now on, I'll follow your advice and try to keep them apart when his father is home.'

'Kendall is becoming an introvert. I fear for his future at home, unless you send him to a reliable college in Australia. Parents pay fees for children to attend public schools here and in England. In Australia the government contributes to their books or excursions. Your son needs children his own age to play with. His intellectual knowledge is astounding for a child of four. Please *Ma Cherie* keep in touch with me on what you decide.'

She didn't regret talking to Pierre. It solved a lot of difficult problems and they had arrived at an amicable conclusion. Gigi knew she and her family must relocate to a warmer climate and she valued his friendship.

She admitted to Pierre she would willingly confront her husband on her arrival in Evesham that evening. Locking their cases his words came flooding back: "Kendall is becoming an introvert".

Sometimes he flew into a temper for no apparent reason. She wouldn't let her son emulate his father's dictatorial traits and brutal behaviour. Gigi knew his horrid attitude to Kendall would ruin his gentle character that she'd spent years in cultivating. Nor would he let Kendall attend a public school. Children carried tales and caused trouble for their elders. Ian couldn't afford a scandal to emerge from gossip. He held a grave suspicion that he may still be on the wanted list of Nazis who'd committed atrocities in the last war.

Arriving home before her husband, Gigi wept with relief. Her anxiousness had lost its momentum with the rising moon.

'I'm glad we spent a wonderful week with Pierre in Paris. They're a delightful family and I appreciate their honesty and yours, Kirsty. Now I can now face Ian with an easy conscience. I'll feel my way until he's in a receptive mood, and I won't tell him of our prospects of leaving for Australia until he is sober. When he's drunk or irritable he swears and is more belligerent. Have a safe trip. What time does your flight leave Heathrow in the morning? I won't say farewell. It sounds so final.'

'Au revoir my dear, is not goodbye. God willing we'll see you all in Australia before Christmas. Under your night lamp is a little gift. Every time you use the contents Gigi, think of me working under a hot summer sun, in the field of sheep husbandry.'

They joked of long hours they'd spent together in their youth in Paris. And of their trips when they travelled with Pierre through the provinces of France.

She kissed Gigi's cheek at the door. Tears gathered in their eyes on parting. Kirsten had decided to leave before Ian came home. She couldn't afford to be the instigator of another argument between the couple. If he even suspected that she had been in their home, or even in Evesham it would inflame his temper and the situation. Thinking, her perspective senses were aroused. *In retaliation, he would have great pleasure in denouncing Bjorn's promise of a job. Then Gigi won't get the chance to put our proposal calmly to him. She repudiated every suggestion I put forward and ignored Pierre's plea for her to accept our offer to be their family sponsors.*

Kirsten missed not giving her godson a cuddle. She kissed her hand and placed it on Kendall's forehead. *The dear boy looks so pathetic lying in his cot: His skin looks puny for a child nearing five. I'll never forget how he giggled when Gigi put him down not an hour ago. If he awakens and cries, I might be tempted to stay in Evesham. That idea is too ridiculous to contemplate. It'd be a disaster if his father arrived home unexpectedly.*

Checking the time, she realised the hire car was due to return from a shopping excursion in the village. Waving goodbye Kirsten hurried to the idling vehicle.

Gigi's thoughts travelled on a parallel level. She now realised Pierre's advice was genuine. Within a passive mind words began to form how to broach the subject of leaving England to her husband. In a serious mode, she began to imagine how Ian might react to the Svensson's idea for them to emigrate.

In a reflective, contented mood she prepared her son's evening meal. Kendall was overtired and irritable on awakening. Settled again, the peaceful silence of a passive mind encouraged her to rest, not eat. Within minutes she heard the hum of a car's motor idle as it cruised to stop outside their cottage. When the motor roared to life again she thought, *No, it's not Ian thank goodness. I'll have time to enjoy this hot cocoa before he awakens.*

She took the cup and clean clothes to the bathroom. 'This hot bath is relaxing.' Letting the water trickle down her back she sighed with relief. Towelling herself dry, she dressed in a powder-blue woollen frock. Like her, its silk-embroidered yoke had seen better days.

Kirsten's lavender *eau-de-cologne* heightened a euphoric mood. Underneath the atomizer sat a note which read: "Darling, use this perfume sparingly. Our plan is to make the honey pot attractive. Not frighten the 'B' away".

Gigi sprayed a little perfume on both her wrists. Re-reading the note, with a hint of sensual aroma on its paper, she smiled while destroying it. A light blush of pink rouge accentuated her sunken cheeks. A slight glow shone through the translucent powder on her high-cheek bones. A fine spray of perfume on her hair was ample to give the soft filaments a glossy appearance. Her fingernails, brushed with petal-pink polish glistened in the subtle twilight.

Rummaging through an old trunk she found two tortoiseshell combs. She recalled having seen them in a drawer at the Edelweiss Pension in Switzerland. *Ian must have put them in his travel case. I didn't.* She held them against her hair. *They aren't my initials. M.K means nothing to me. Was the woman who wore the combs beautiful?* Fondling each comb, she pondered over their beauty and where to place them.

'Should I wear these? Yes, they'll keep my hair neat and hold the wispy bit tidy. The honey-gold shell will highlight my dark hair and my amber pendant. Pierre's gift to me in Paris. He's such a dear. I love him dearly and treasure his friendship.' She held the combs in various positions. Deliberating, she tucked one comb under each side of the French roll. *Now they look perfect. Ian probably won't notice them, or me.* She flexed her gaunt shoulders. *Never does these days.*

Muttering to herself had become a favourite pastime. With only her voice for company it preserved her sanity. Although an intelligent child, Kendall couldn't carry on an adult conversation to bolster her spirits. He boosted her morale and life. She doted on her son. His little arms embraced her on awakening. Even in the land of dreams, she must have come to mind, because he often grinned and called her name in his sleep.

Never would she give into his tantrums. If he misbehaved or gave her cheek he got a smacked bottom. His father only put food in their mouths. He never taught Kendall a thing, except how to exhibit an irascible temper. Not a charitable example for his child to follow. Reminiscing kept her mind off miserable incidents while bathing Kendall.

Determined her son would grow up to respect his elders Gigi sighed.

'He's of an age now where manners and politeness count.' Sipping the cold cocoa, she recalled his recent bout of ill-temper. 'Please instill in my boy the wisdom of youth, the kindness of age, the knowledge to distinguish between right and wrong, truth and honour. My plea is for you to grant me all that my heart desires, dear Lord.' This short prayer her grandmother had prayed with her as a child, each night before they retired.

11

Ian thundered in right on teatime; threw his coat over the arm of the lounge-chair and pushed past his wife to wash before eating. His hat sat askew on a coat hook along with a woollen scarf, still warm from hugging a drunken neck on the brolly stand.

A harsh voice echoed from the bathroom. 'So you do know where you belong?' Sarcasm was rife in the deep-throated timbre. Pleased with himself, he concluded in the same acidic tone. 'Don't expect to go again, I won't let you. I hate cooking for myself. A man must admit whatever is brewing on the stove smells delicious.'

'Your favourite, steak and kidney pudding,' his wife responded timidly. 'Kirst...' Gigi hesitated in saying the wrong word. 'I bought the meat on our way home from London and thought you'd enjoy a hot meal on a cold night.'

Her diplomacy worked. His anger diminished instantly. With only a minor sting in the tail, his voice echoed in continuum. 'It's good you're home.' He smirked, standing by the kitchen door. 'At least you remembered that I still exist.' Wiping his fingers on a hand towel, he dropped it on the floor. *Why should I worry, my slave will retrieve it.*

Rather than cause a row she swooped on the hand towel. Arguments were non-existent this evening. Not in the good mood he seemed to be displaying.

'My slippers are on the hearth and those cinders have a healthy glow. At long last you're considering me with respect. To what do I owe the pleasure?'

'I always warm your slippers, the moment I hear your car.'

Impatient, his eyes expressed anger. 'Dish my dinner up now. I'm hungry. And stop snivelling, Gigi. Where's Kendall? Running the streets, I suppose.'

'Ian, our son is asleep in his warm bed. He's a very tired little fellow…' She almost said the wrong thing again. *Ian seems in a reasonable mood. I think he'll accept their offer, if I put it to him calmly. I hope he doesn't go ballistic, like he often says I do.*

A mistake of this magnitude would inflame his temper. Then she would leave herself open to ridicule or abuse. Gigi needed him to be in a passive frame of mind tonight.

'Kendall went down early, so we could have time to ourselves. I appreciate all you do for me, Ian.' *Buttering him up with gentle platitudes will make him receptive to their suggestion.* It lay heavily on her mind to say: *What would you care, if I let him run wild in the street. You never have worried until now.* Instead she refrained.

Dinner was quickly demolished. Ian went one better than Oliver. In an exuberant mood, he ordered a third portion. Turning to his wife, now standing by the sink, he snarled in an over-polite affirmation.

'That meal was delicious. Why can't you cook a decent meal every night?' He licked his jowls, like a mutt would do. 'I have several items to talk over with you, not at the table. Brew my coffee while I unpack. Then I'll tell you of my news.'

This complacent change in attitude wasn't conducive of his brusque temperament. 'What is he conjuring up now? Why is he so amiable this evening?' Gigi uttered, putting the kettle on. Then another idea sprang to life. *He might be sexually aroused. He never compliments me. Not without some ulterior motive lurking behind his insincere flattery.*

Criticism and Ian Ross were inseparable. She decided to be composed. By keeping calm, she could deliver the ultimatum her mind had trampled to threads.

Comfortable on the lounge he gestured for her sit alongside him. Without warning his fingers travelled up her thin legs, no longer shapely.

'I'm feeling amorous tonight. After four years of wedded bliss, you haven't lost your innocence.' Slowly his fingers wormed their way to her thighs. 'I did miss you Gigi, while you were mixing with the social circle of London.' Salaciously surveying her cleavage, it stirred to life the masculine demon within his loins. Every heave of her emaciated breasts excited him and heightened his libido to the pinnacle of ecstasy.

In a desperate ploy for peace, she pushed his fingers away. A hard pinch on the tender flesh of his wrist made him withdraw them.

'Need you be so cruel? That hurt me! Gigi, I can't afford to be caught again. When is your next menstrual cycle due?' His hand roamed around her tummy, under the skirt of her threadbare frock. 'My dear, has the cat bitten off your tongue?'

'Now is safe. The… there's something I need to ask you,' she stammered. 'I need to know if you'll leave for Australia with me. Kendall can go to school there and we will be sponsored …' Her courage waned. *Dare I tell him the Svenssons have offered to be our sponsors?* Fearing some form of reprisal, she cringed. *If he gets angry, he'll probably thump me in the face.*

'Out with it, woman. I'm not a mind reader. You interrupted what I wanted to say. Now finish what you meant about someone sponsoring us to Australia.'

'Will you come with me to live there permanently? You'll have a job working on a cattle stud, no a sheep station. The owners will show you how to manage it and pay us both. They've offered to let us live in a house on their property. There, now you know.' A hushed gasp escaped through her closed lips. And she could breathe easier.

'One of Svensson's schemes, no doubt,' he scoffed yet inwardly smouldered. 'I'll need several days to think over their proposal. I wanted to tell you, we're leaving here soon. A good break and warm, dry air might revitalise your hollow cheeks. Their proposal sounds meritorious and a prosperous business venture to undertake.'

'You have twelve hours to make a decision, not days. If you refuse their offer to live in Australia, where else could we live, Ian? You criticise me with little or no reason, if I complain about the bleakness of this house, because we lack coal. You've forbidden me to light the furnace. Every night I shiver and keep Kendall warm by nursing him to sleep. He's wrapped warmly in my topcoat, so his little body won't freeze. I've pleaded with you numerous times to get the roof mended. That loose thatching rattles in the wind. It frightens him and then he cries. You led me to believe that you like living in Evesham. I love walking with Kendall in the Cotswells on warm spring mornings.

'What rot! I've never said that at all,' he snarled. 'You're dreaming. I'm tired of teaching brats in Stratford. Idiots who lack manners. The parents don't give a damn if their kids are insubordinate and threaten to murder their lecturers.' Ian didn't elaborate further. Giving her leg a tap, his lips brushed the wispy curls on her cold forehead.

'I'm going to have a shower. Prepare our bed with the little towels. I hate sleeping on stained sheets. As I told you before, I'm feeling amorous tonight. Do it now, Gigi.'

A short time later she followed him to their room. Hesitant, she seemed reluctant to let him fondle or kiss her nipples. She relented to his gentle foreplay of the delicate flesh, beneath her dark pubic triangle.

6 am Breakfast the following day: Ian condescended to listen to Gigi who relayed the Svensson's proposal to him while Kendall played with a toy truck beside the fire-screen. She explained how they had promised to back them in their new venture. 'Once we learn the ropes, we can buy our own homestead and property.' She had quoted Kirsten's note word-for-word.

A smile looked out of place on Ian's grouchy face. 'You took the words out of my mouth. This once I'll agree with you and them. I knew they were behind this nonsense. Still, the prospect of owning a sizeable sheep station in Australia sounds promising.' Elated he admitted, only to himself, that he knew nothing about running any kind of a station, or a huge property. *I'll allow them to show me how to manage a sheep stud. They better not think I will lower my dignity by dipping sheep in whatever they dunk them in. Crutching a sheep, what a repulsive idea. That I will never do.* Ian gave this considerable thought before admitting, *If I approach this problem sensibly it will diversify my funds. Yes, I'll accept their offer to emigrate. It has merit and could prove profitable in the long run.* Refolding his napkin, he continued to affirm in silence: *If and when I buy my own property, staff shouldn't be a problem to make it a probable investment. If I continue an extravagant lifestyle, my funds will be depleted in five years, maybe less.*

Worry sat heavily on his mind. In the past he'd connived to keep his finances viable. His funds were diminishing at such a rapid rate, Ian realised that unless he recouped the loss he would be bankrupt. An insatiable sex life, not with his wife, would destroy his future. *Without diversifying my funds, I'll lose my most viable asset, her.*

Undesirable memories were haunting him. *My deceased wife's jewels, that valuable oil painting and her legal documents are in my security box at our local bank. I'll go in tomorrow and withdraw everything. It'll be difficult, if not impossible to ascertain what this bitch has leant or given to that French bastard. She doesn't have a bank account. I control our finances. Gigi has an allowance of twenty-five pounds a week. One morning I'll rummage through her personal things. I don't trust her, she can be deceitful at times.*

Meditating, Ian then remembered, *my documents, and her valuable trinkets are in that bank. I'll write a release note. If Gigi doesn't sign it, I'll threaten to put Kendall in a home for orphans or wayward kids in London. Then we'll leave here. In Australia I'll make sure she's has no contact with the brat. Then I'll take great pleasure in burning all correspondence from France in front of her.* Ian proceeded to note down all her bank items, and stuffed the list in his coat pocket.

Four separate incidences in Stratford had greatly perturbed him. They made the fine hairs on the nape of his neck stand on end. It devastated him when confronted with his once proud, Nazi past.

A surreal happenstance warned him to be extra cautious. He suspected the local police were investigating him and the position he held in Wehrmacht. He also suspected a colleague from Germany might be plotting to turn him into the Mossad or MI6. The previous month he narrowly averted bumping into two ex-colleagues in Stratford. And knew it wasn't coincidental.

The third unpredictable incident occurred earlier in the week. Strolling down a narrow lane leading to a tavern Ian often frequented in Stratford, he almost ran headlong into a couple of Scottish friends. Professor Hadrill, one of his ex-tutors in Edinburgh University had turned to speak to his son, Richard. Initially Ian imagined he may have recognised him. Evidently not! He'd darted down a side lane. Neither he nor his father followed him, a man they knew as Rolf Breusch. Nor did they know his Germanic surname of *von* Breusch.

The fourth incident occurred one evening, prior to Ian returning home to Evesham. He and a lady friend were dining in a local restaurant. At a next table were a colonel and his wife; both of whom he'd recognised. On reflection, Ian knew they were acquainted with his first wife's cousin.

When the colonel's wife left to visit the powder room, he approached Ian. 'Well I never, you're von Breusch. It's been over two decades since our last meeting. From what I heard in the Chancellery, you were supposed to have committed suicide in Berlin.' The German ceased speaking and looked startled at Ian Ross.

'Sir, I'm afraid you are mistaken. Damn common face I must have. You sir, are the third person who's made that mistake this week. Damned annoying, what! Richard Haney's my moniker,' Ian spoke in a posh tone, which accentuated his English accent. The name he'd plucked from browsing through a novel on a bookshelf in this café.

'Sorry! I could've sworn you were Rolf von Breusch…'

'Never bet on uncertainly, old fellow,' Rolf, under his assumed name of Ian Ross had laughed and it sounded forced, but real. 'It'll break your pocket. Mistakes have a habit of reoccurring. Many men have lost their strides for less. Enjoy a beer on me.' Ian had slammed two English pound notes on the table. 'Whatever you do sir, do *not* interrupt us again. I'm about to ask my lady a poignant question …'

This pause was deliberate. He suggested they should leave the restaurant. Escorting the prostitute down a flight of steps, Ian left the bewildered officer to mull over his last comment. Adding insult to a string of lies, he glanced back at him and his anger subsided.

As we walked past his table I recall saying: "A man can make a mistake once. Only a fool makes one twice". His response to my snigger still lingers on my mind: I felt his glower following me to the car … that night. It's pleasurable to think, the bastard will never inflict his personage on me again.

Consoled in his own world of past infidelities, Ian failed to notice his wife's bemused expression. Reminiscing, he didn't hear Gigi speak, while clearing away the breakfast dishes.

'Ian, you're very quiet. Am I the cause of your anger? I do my best for you and you never apologise. Why! Do you think I'm your servant, or expect me to cower every time you roar at Kendall or bellow at me? You're never grateful.' She was mystified by his vacant expression, panting breaths and sweating brow.

'No, it's not you. I was mulling over the idea you mentioned last night. Migrating to Australia sounds interesting and it has merit.' Patting his mouth with a serviette, his brow furrowed. 'What's the best time to contact the Svenssons? I notice the phone has been reconnected. You must have some idea when and where to call them, Gigi.'

Finding it hard to think with Kendall whining she didn't reply immediately. 'You promised to let me know your decision this morning. Am I to take it that you mean, yes?'

'Of course.' The response was equally as terse and his manner disconcerting. 'I have given their offer intense consideration. Let me finish speaking before you argue. There are conditions! Your French friends will not be invited to our home. Nor will you respond to their letters. I demand privacy, whether it's on the Svensson's property or my own, if I purchase one. My third condition is Kendall must attend boarding school in Australia. You can't expect a man to learn "the ropes", as you tritely said last night, with a brat underfoot. I gather you imitated their words. It's not a term I'd think of using.'

'Ian, don't speak of our son in that tawdry manner. He's a four-year-old child. Not an adult. Don't send him away please. I couldn't bear living in Australia without Kendall,' Her gasp of alarm sounded deafening when he raised his hand. She knew why.

'Shut your mouth, or I'll shut it permanently.' He looked at the thin-bladed boning knife, then at his battered razor-strop hanging above the sink. 'You leave me no option than to enroll him here and we'll leave London. It's your choice. Now, where can I contact the damn Svenssons? I'll phone them, now. Not you.'

Insulted by his supercilious snarl, she let the porridge pot fall. Hot sludge doused her worn slippers. Gigi responded to his insufferable remarks with a grunt. In an effort to not antagonise Ian, she lifted Kendall out of the playpen. Cuddling the child, she responded in a subdued tone. 'Kirsten won't be home in Australia for another week. I have her hotel number in London. Around teatime would be the best time to phone her.' Gigi tried to think what time Kirsten had stipulated. Confused, she couldn't remember.

'Forget it! You're a useless twit. I'll ring her now. I need a reply, before I can make arrangements for that brat's schooling.' An insensitive father, Ian glowered at his son, whose little hand clasped a toy truck. Kendall had fallen asleep on his mother's shoulder.

'Put him to bed this instant. Then come to our room. You enjoyed our sexual liaison last night. Now I want to celebrate our change of fortune in peace.'

'Ian, it's barely nine in the morning.'

'I don't care if it's midnight. If I want sex I'll have it, irrespective of the time of day and without your consent. Put him to bed. How can a man think of making passionate love to his wife, with a squawking brat disrupting his pleasure?'

His demand she took as a threat. *If I refuse him again, Ian might retaliate by smothering Kendall with a pillow, like he tried to do a month ago. That same morning, he cornered him in the toilet. I intervened on hearing his screams. Ian was trying to force his little head under the water. I wish he'd stop drinking to excess when he's angry. In such an aggressive mood, he could thrash his little bottom with his belt. If I interfere, I'll get another smack across the mouth.* On her way to settle Kendall down she looked in the hall mirror. 'My lips are still swollen after our argument in bed around dawn this morning. His breath then smelled of garlic and booze.'

The previous week, in a drunken rage Ian had tried to drown his son in a bath. Only her quick thinking had saved Kendall's life, after she found him missing from his bed.

Feeling extreme discomfort after sexual intercourse this morning, while dressing in a blue vyella blouse and woollen skirt, her voice lowered to a whisper. 'I know Ian is capable of murder and he must never be trusted or underestimated. While Kendall remains with me, he'll be protected to a degree. Before leaving Orly Airport in Paris, I promised Pierre, who mistrusts Ian. I would never leave Kendall alone with his father. Have I failed?'

With his sexual appetite appeased and Kendall asleep, Ian penned his business letters in the small snug he called the den.

Worn to a frazzle from arguing over their new adventure, Gigi finished the ironing. Thinking of their decision to emigrate, she wondered how they would manage in Australia. Neither she nor Ian knew anything of the huge project they were about to undertake. It would be a massive intrusion in their lives. Intensively cajoling over a cup of cocoa she came to the following conclusion: This chance to lead normal lives is imperative. With Kendall at school I can enroll in a college to study economics. With Kirsten's help, I should master the art of bookkeeping.

With no alternative, Gigi accepted the fact that Kendall would be enrolled in an Australian boarding school in the country town of Yass in New South Wales. She surmised that being separated both their lives would be miserable.

There were no close ties now since her grandmother's death to prevent their family from leaving England. It appeared Ian's lie of his aged mother dying in Scotland conflicted with what she suspected to be the truth. She believed everything he told her. At a loss to understand why his lucrative position up in Stratford had been terminated, she confronted Ian on his return home from the local barbershop.

'You haven't told me the real reason why they fired you. It seems so sudden. Are the Stratford police investigating some misdemeanour, they *think* you may've done in that private school? You've been teaching there for more than twelve months. I'm confused. You seldom tell me the truth.' It lingered on her mind to ask, if he'd abused a young student. *No, he would be foolish to sexually approach a young girl. Nor would he risk molesting a child. There must be another reason for his sudden dismissal.*

Scrubbing the shower recess she uttered, 'Perhaps they dispensed with his services because they have less students. Something serious must've occurred for the school principal to have sacked him. I don't know what made Ian change his mind about Australia. Neither of us know anything of breeding sheep. We should give their idea a lot of thought before we sign a document. In time he may treat Pierre with respect and look at things from a different perspective. Ian can't prevent me from contacting his family. I might phone the Svenssons at their home. I think she called it Jalna. Tomorrow morning after Ian leaves to find work, I'll a jot a line to Kirsten. If I do it now, the postman will collect it on his rounds. He arrives here around ten.'

Dreaming of her Parisienne friends while mopping the floor, Gigi's smile faded to an obscure frown when she heard the study door slam. *Pierre and his family will miss Kendall, more than they will me. And I'll miss Mercedes telling him stories of historical France before the last war. His cute laugh is still ringing in my ears. I better pop in and see if he's stirred. Ian banged the den door a moment ago. While he's busy, I'll give Pierre a tinkle. He should be home from the office at this hour.*

Wistfully she crept into Kendall's room, then through to the master bedroom. Closing the door, it eliminated the risk of Ian hearing her on the phone. Her call to Paris was answered by Pierre in his conservatory.

'Darling, I can't speak loudly. I'll tell you what I need to say. Please don't say a word until I finished.' While he acknowledged her plea, Gigi listened. With excitement building to a crescendo a feeling of terror overwhelmed her. Aware Ian could barge in unannounced, she held her breath. 'Yes, he knows everything. No darling he didn't object and has accepted their offer to sponsor us. He has one objection. If we go to Australia, he insists on Kendall attending a private school. We know the sanction and our visas will take time. Government red-tape ties up most projects. Pierre, I'm sure Ian will keep his promise, now he knows the truth. Tell Kirsty when she rings in an hour, Ian will call her in London around seven tonight, British time.'

Gigi paused to hear what Pierre was trying to say in English. 'Please ask her to stay close to the phone. Yes, I did get a shock when he agreed to their proposal. A superb move and I hope it works okay. My love to you Pierre and your family. Bye darling.'

This ordeal over, she replaced the phone and then jumped on the bed. Emotional tears of joy flowed. Her exuberant sighs outlasted the tears. They emulated her heartbeats. Ecstatic with excitement Gigi clutched her breasts. Years of agony and frustration were released in a sparse few minutes and the severe fluttering of her heart was diminishing.

Putting a warm coat and her yard shoes on she strolled through the wilderness that once was a beautiful garden. This evening the dead rambling roses and faded lilacs seemed to exude an array of sensual perfumes that heighten her awakening senses. Her pent-up anger gradually ebbed away with each breath. Hope for an unknown future began to blossom in an overstressed mind. Humming a mythical tune Gigi walked on a cloud of positive expectations.

Kirsten would be awaiting Ian's call. This alleviated her anxiety. It placed less strain on both her and Kendall. Provided her husband didn't run into hassles getting through to London, a positive frame of mind would boost his morale and may even sweeten his sour temperament. The thrill of escaping a bleak English winter and resettlement in a new country encouraged her to smile. She couldn't foresee any reason why their family wouldn't be accepted as permanent migrants. Gigi had heard about families from the British Isles being sponsored to Canada, New Zealand and Australia. In a spritely mood, she heard an angelic voice whisper to the last flush of autumn leaves: 'Never fear little mother, you and your son will sail on a ship across many seas to the land of your dreams.' This gentle sonnet revived a cord of remembrance. *Every day I feel as if Manny is watching over me and walking beside me. An angel on my shoulder. I wish her could be with me in our new home in the land of kangaroos, native animals and wild flowers. Still, she will be in my mind.*

Twelve months later the Ross family was accepted as migrants. With limited time at their disposal, Gigi rushed to pack their personal things and catch the train from Oxford down to Southampton where they were to board the ship bound for Australia.

They managed with a day to spare. Gigi took Kendall walking through arcades and other venues in and around town, whereas Ian mercifully managed to refrain from imbibing. This enormous feat proved invaluable to him and a bonus to her. Excited about his first sea voyage, which looked decidedly promising, Kendall waved streamers as the fore and aft tugs edged the ship away from Southampton's wharf.

On board, Ian's left foot braced the brass rail beneath the top deck rail. Impatient to be underway he said something harsh to his wife. Scowling she held Kendall's hand and together they moved further along the deck.

Eager to unwrap Mercedes and Pierre's gifts for his fifth birthday, he spied his mother weeping. Perplexed the boy asked, 'Mummy did Daddy get cranky with you? I saw him give you a nasty look.' Snuggled between her skirts for protection he shuddered in fear. 'He won't beat me again, will he? I have been a good boy all day.'

Kendall climbed on the narrow rail beside his mother who cuddled him. 'Darling boy, I promise that your father will never hit you again.' She showed him the subpoena in her hand. 'This letter says your papa must behave or the police will take him away. Remember papa promised not to hurt you, so you're safe here with me.' She shivered as winter's bitter wind whistled around their legs. Kendall gasped with glee on seeing his streamers snap as four tugs pulled the ship into midstream.

With eyes brimming Gigi waved to people wharf-side who responded in kind, blowing kisses to their departing friends. She wished Manny could be here to wave them farewell. Instead, she enthusiastically trilled to seagulls high on the wing. 'Australia, here we come.' Her voice dropped an octave as tear-sodden eyes focused on busy faces far below on the wharf. 'This sad parting ... we will remember son. Darling, I know you'll miss seeing your beloved Papa Pierre and his family.' Silently she sighed. *I will dreadfully.* Still, time in her wisdom has a way of mending friendships, broken but not forgotten.

The boy nestled into his mother's arms, as she reminisced of Paris glittering under a trillion of summer stars. This romantic vision of a dream faded to a memory that would forever linger on her mind.

Frightened by the three sharp booms belching forth from the ship's funnel Kendall jumped and cried to be taken below decks. In their cabin he peeped through the porthole until their ship turned downstream. It then steamed for the Isle of Wight. Their journey to new life in a strange land was beginning. It took six long weeks of ploughing through white-capped waves on rough seas before the ship docked in Melbourne, Australia.

12

*1*960. Since the spring lambing began on Jalna none of the local or
drifting shearers could afford to relax. Today at six, on conclusion
of crutching and drenching thousands of his sheep, Bjorn decided the
fellers deserved a spell. Sheltering out of the impenetrable heat BJ and his
foreman stood under the drooping leaves of a ghost gum. Pensive, they
observed the ringers begin a breaking-in procedure. The tree's spreading
boughs provided inadequate shade to protect them from intense heat
rising from the hot parched earth. Even this early in the morning
perspiration poured down their shoulders in a steady stream. Sweat-rags
round their necks were useless to stem the constant flow. Sour-stinking
mess saturated the men's navy cotton singlets that clung to the weathered
skin on their backs.

Most of the men standing around the compound were naked from
the waist up. Having discarded their shirts before an unsympathetic sun
had gathered strength to scorch the earth, furrowed and contoured to
parchment. Shirts would soon be donned once the searing rays bit into
their flesh. Even Jalna's four working kelpies panted while lolling under a
peppercorn tree near the fence.

'Just as well we began this job in the cooler hours,' Bjorn remarked to
his under- manager who, for unexplained reasons, ignored the comment.
Breaking-in was one job every wrangler on the property wanted to
participate in, or observe.

Aware the day would be a sizzler BJ requested his workers to get
stuck into their work before the sun climbed high above the woolsheds.
Breaking-in his young stallions was terrific training, it taught the up-and-
coming wranglers the moves to make while still remaining on their feet.

Casper, a fifteen-hand stallion was stubborn. This feisty beast would soon knuckle down to Big Wally's commands. Jalna's top wrangler was adamant by smoko Casper would relent his hold of trampling the steamy patch of raw earth. Wally's determination sweltered on the horse becoming subservient to his will and it would soon learn what each tug on the reins meant. Clinking of the bit between his strong teeth was yet to be tried and tested. This workout must be mastered before the hot sun burnt off black clouds brimming the western sky.

Jalna's chief breaker Wally Hall, the men had nicknamed Big Wally, nudged his boss prior to breaking of this horse.

'Boss, I ain't gonna let that bay kick me butt in the dirt. He's not going to get the better of me. I'll bring Casper down to his knees first.' This man with a strapping build respected all the horses he worked with or broke-in. Take their nonsense never, and they knew he was the master.

This bloke, reaching six feet tall, had a different opinion of the border collies and red setters. Those mutts were either good rounders or dead meat. Wally wouldn't feed some of the critters he'd worked with in his time. His boss of the day and his companions were set straight on this score well before they began the work.

Ian Ross leant against the compound fence. One boot rested on its lower rail. With his bronzed elbows bent, they rested against the top fence rail. His sauntering arms dangled at will and one hand clutched his hat. The air of obstinacy in his manner defied BJ's previous warning. A deep frown momentarily disfigured Ian's suntanned features as he gazed into the strong glare. And his hatless head caught the boss's attention. Bjorn noticed some of the other men had removed their slouch hats. Rather than tackle Ian, his under-foreman, he roared at the roughies whose hands were also bracing the fence. Waving his hat, he pointed to each man in turn.

'Where are those damn hats of yours? Either shove them back on pronto, or have your pay docked. You lot know my ruling. And don't forget it.'

Instantly a fanfare of headgear was smartly donned. Bjorn wasn't going to allow his crew to take time off work with sunburn, because they'd stupidly flouted the bushman's law. The older men knew better and weren't game to go without their all weathers. Every ringer knew from experience how a man would fry under a sweltering sun, even on a cloudy day. They also knew the heat of a Yass summer would cook a bloke and fry an egg within two minutes.

Besides, no man in his right mind would allow the boss to dock him a day's wage. To their way of summing up, they worked bloody hard for their dough and wouldn't part with it from choice. Only the Missus, pub, or perhaps their horses had that prerogative.

Bjorn Svensson worked the best crew this side of the range. In return all the blokes respected his judgement. All except one! This particular individual resented being told what to do, in the way of running a station. If orders came his way, Ian Ross would shoot through, back to his and Gigi's bunkhouse.

The bunkhouses on Jalna were quite comfortable and spacious. The Ross domicile consisted of three airy bedrooms, a washhouse plus a private bathroom and an outhouse. The toilet, or dunny as it was commonly called on most Australian homesteads was located in a separate block not ten paces from their door.

Their hut's self-contained kitchen was immaculate and clean. Combustion heating fed through pipes from the wood-fueled stove provided heat in winter or on bleak days. Ceiling fans allowed its occupants to cope with the extreme heat of summer. All modern appliances were provided. Jalna's supply of power was automatically boosted by four generators in horrendous storms or if the town electricity grid failed.

Three earthen dams, all a reasonable size, provided enough water for the herd of a hundred Herefords and a mob of two thousand sheep to keep from dehydrating on insufferably sweltering days. Cattle often cooled off in shallows of natural water-courses. Artesian bores in Jalna's top paddock were backups for the dams. Three huge cement tanks allowed its owners, staff and guests ample rain water for personal or household use. The original property owner had made the entire homestead and lodgings self-sufficient prior to his nephew taking over.

Sam Svensson's homestead and land holdings were still at Young, a sizeable town around seventy miles north-west of Yass. Sam remained Jalna's financial backer. He helped the Svenssons with difficulties they struck regarding droughts, or loss of sheep or lack of fodder. Once the property was up and running to his satisfaction Sam pulled out, letting Bjorn manage his own stock and feed lots. Funds were available if either he or Kirsten hit trouble or needed extra money to bolster their lucrative business.

Together the Svenssons managed to run their sheep station to perfection with hand-picked workers, most of whom had stayed on after they took over Jalna. The entire staff thought the sun rose and set in BJ's pocket. Kirsty, the bookkeeper, kept their finances under control. Jalna's prosperity rode on the sheep's back.

Most of the workmen placed themselves at Kirsty's disposal and would do anything to help her with little problems. BJ let one of his reliable men drive Kirsty to town if her vehicle was off the road or in for service.

This luxury was never afforded to Ian Ross. Neither Bjorn nor his manager trusted Ian with women. Over recent months, when he and Gigi's backs were turned, his play for Mrs Svensson's attention had been blatantly obvious. Kirsten complained bitterly about Ian entering the homestead unannounced. Before breakfast while dressing after a shower she caught Ian observing her. Extremely embarrassed when he licked his lips, her foot had sprung into action. The bedroom door slammed in his face.

Safeguarding her boss's privacy the housekeeper, Mrs Raddick walked up the main hall and stood near Kirsten's bedroom, especially if she noticed Ian on the prowl.

Helen Raddick, a forthright woman in her mid to late fifties, mistrusted Ian Ross with their housemaids. Jalna's cook, she ran rings around local women with her culinary talents and often won prizes at both the "Coota" and Yass shows. Once a week she would drive down to Canberra or into Cootamundra and took Mrs Ross along for company. This allowed Gigi space to breathe without the indignity of being manhandled by an angry husband when he was in a savage mood. The ladies in their social group considered Mrs Ross a superb cook. Even the colleagues in their equestrian team were conversant with her culinary expertise.

Socialising with these women caused numerous arguments between her and Ian. He objected to his wife having the luxury of freedom while he worked. Catty innuendos would be cast regarding her attention to their men. Everyone knew the accusations were groundless. If BJ heard them, Ian would cop the rough end of his tongue, or feel his boot in a very delicate part of his anatomy.

Gigi reveled in this relaxed country atmosphere and respected the local inhabitants. Her kindness was repaid tenfold by the workmen who embraced her with endearments. The majority of the male staff idolised the tiny Welsh dancer.

Since she'd begun working on Jalna the gauntness in her face had dissipated. And the thinness of her body no longer existed. She embraced life with sparkling eyes and vitality. Not the life she refused to remember. It belonged to the past, not to a future filled with grand expectations for her son.

The vivacious spirit she portrayed captivated the roughest diamond on Jalna. The neighbouring homestead's "top-gun" Lanky Hancock, on loan to Jalna shearing's shed, was her most ardent admirer. An old woolly from way back, he thought, when she was born they must've discarded the mould. No other woman possessed her keen wit and bonza humour. In his estimation Mrs Ross was a good looker, cos no one could match *his* special *girl's* charm. Gigi enjoyed the camaraderie between him and Bjorn's men.

In the off-season Lanky worked younger members of Jalna's crew. He took pride in teaching the lads how to ride with the milking Herefords. A smaller herd of fifty "killer" cattle were bred solely to stock the cool rooms with well-hung beef. Sheep provided mutton and hogget. By-weekly sides of pork and boxes of cured bacon and fresh farm produce plus eggs were delivered by the piggery and local growers then stored in Jalna's walk-in freezers.

Their nearest neighbour, other than the property adjacent to Jalna, was fifteen miles north from Abergeldie's property. This distance wasn't unusual in country areas. The small township of Yass spread in all directions, well beyond the district's southern pastoral boundaries.

A couple of times the stockmen had warned Lanky that unless he wanted a clobbered head, he should keep his distance from Mrs Ross when her old man was around hearing. Most of the men despised Ian over his jealously. None of the staff trusted him.

Lanky Hancock consistently threatened to tackle Ian over the way he spoke to his wife. At loggerheads on many occasions they were separated by the young boxer Corey Sidle or his friends. This problem was soon resolved by Bjorn keeping them apart. It didn't prevent Lanky from doffing a tattered Akubra whenever she passed him. He would present her with bunches of boronia or wild flowers gathered from the top paddocks. Lanky never plucked roses from the miss's garden. That would set the possums in a flurry and cause dissention in camp.

During their morning smoko the men topped their canteens with fresh, cool water to sip and to replenish their fluid intake after excessive sweating on the hot summer days. Mrs Raddick carried out cakes and biscuits she and Mrs Ross had cooked before the real heat of day rode in. Helen put the tray on a metre high disused cable reel. When up-ended the wooden spool worked as an ideal table. Flat, it couldn't topple, the base being a similar diameter to its top.

Interested in this procedure Gigi stood inside the screen door and, recuperating after a fall, she was limited to doing light work in the kitchen. With a pronounced limp still evident she wasn't allowed to tread on uneven ground or do heavy chores. Kirsty or Mrs Raddick offered to drive her to the doctor's surgery in Yass again in two weeks. Gigi's supposed "accident" a week earlier still caused her immense pain in the left hip and corresponding shoulder region.

She was fretting after Ian had packed Kendall off to boarding school in Goulburn without a word. The fourteen-year-old boy had copped a smash across the mouth for backchatting his father after he blatantly refused to leave his mother. This hadn't worried Ian whose boot sent Kendall flying against the packed car. On awakening that morning she had searched the bunk and outhouses without success. A note under Jalna's back wire door read, "I've warned you repeatedly. Now you and the Svensson's will have something to really gripe about, you bitch." Even though the note wasn't signed, Gigi had recognised her husband's indiscrete scribble.

After the couple's latest debacle Kirsty thought it prudent for her to sleep in one of the homestead's spare bedrooms. She refused to let Ian come anywhere near the house until after his wife's forthcoming visit to the doctor. Not taken in by Gigi's lies, Stephen Jarvis referred her to a specialist in Canberra for further tests.

On her initial visit, the specialist suspected Mrs Ross had sustained a ruptured spleen and bruised kidneys. Extensive tests proved the damage was from a hard blow to her lumber region which had caused massive internal bruising. The x-ray showed two fractured ribs.

In an effort to conceal her embarrassment, Gigi made everyone think her injuries were sustained by falling down the bunkhouse steps.

The Svenssons and both her doctors sensed the truth behind her injuries were the vicious force of a strong fist or a flying boot. Deep bruising on her chest, neck and facial contusions stank of Ian's handiwork, Bjorn's interpretation of her so-called accident.

Ian's version conflicted with his wife's description of her injuries. Fabrications of her *accident* were deliberate lies. His boot had caused her injuries. The blood on their steel toe caps proved this. Terrified of her husband, she wouldn't admit Ian was the culprit.

Bjorn took the opportunity to challenge Ian while both their wives were in Canberra. Ian's response could be heard at the top of their drive.

'Listen to me, you bastard. I found my wife dazed after she'd fallen in the cool room. That's all I know about her condition. How dare you blame me? I had nothing to do with her fucking accident. I shouldn't have to tell you how clumsy she can be at times.'

'Tell that to the police. Scum like you won't admit bashing a woman senseless. You haven't fooled me Ian. Watch your back in future, because I won't be far behind you. Let me assure you of that. You'll pay all her specialist's bills, and don't think you'll get off lightly. Not with me watching your every move, sport. I forgot, you're not one, are you?'

On his way out of the homestead, Ian turned when Bjorn coughed. 'If you take your anger out on Gigi when they come home, I'll shoot you with this pistol. Not kill you. Then I'll convince the police that I shot you in self-defense after you attacked me. They'll believe my version over yours, Ross.'

Ian didn't bother to reply, just hightailed it back to their bunkhouse to imbibe on cheap whiskey or gin.

The Canberra specialist advised Gigi not to do strenuous lifting or hard work for at least a month. Their return journey home was hazardous in drenching rain. Six miles from the property Kirsten missed colliding with a truck by inches. Shaken, she pulled over on the Southern Highway. After a minute's rest, they resumed their trip home to Yass.

Neither woman relished food. After a hot shower, Gigi fell asleep the second her head and shoulders hit the pillow. Kirsten remained awake to inscribe the day's misadventures and the specialist's report of Gigi's injuries in her own diary. She never missed the chance to itemise every minor detail of their guest's manhandling by a maniacal husband. 'Reports of this nature will be the downfall of Ian David Ross. A beastly man whom Bjorn and I both detest and dislike immensely.'

13

6 am. This morning her first task was to go through Jalna's huge pile of neglected mail and to catch up on lost hours of bookkeeping. After a day's absence bills and belated accounts had almost swamped Kirsten's office.

In the process of teaching Gigi the fundamentals of bookwork, she found her a keen observer and eager to learn. Constant work required keeping every schedule and Jalna's correspondence up-to-date that kept the homestead and its holdings viable. Kirsten agreed with her husband who thought Gigi was a natural with paperwork and a whiz with figures.

This new venture excited Gigi. It occupied her mind and kept unsteady hands busy. She mulled over and regretted Kendall's long absences. Ian's insistence that their son must remain at boarding school affected her deeply. From the moment of their arrival in Australia, a decade earlier, Gigi had seldom seen or spoken to her son. Constantly shifting from school to school in various locations far from home, his father's recent aggressive behaviour proved a blessing in disguise.

Whenever his son came home on school holidays Ian's belligerent aggressiveness increased dramatically. He hated Kendall and the feeling was mutual. Her husband's jealous narcissistic streak she couldn't accept. Gigi idolised her son and she couldn't understand how Ian's spiteful nature was affecting him. It would've been understandable if she were interested in other men. This idea was absurd. A fulltime-boarder in a Catholic college, fifty miles east in Goulburn, she missed her son. Kendall meant everything to her.

Two of Jalna's housemaids carried a tray each of fresh scones and mugs of hot tea out to the shearers, while Kirsten collated their pay.

Having completed the pay sheets, she nodded to her pupil. 'It's time to have our break while the men are enjoying their smoko. A few minutes respite slaving over these accounts won't hurt. They can…' The phone shattered their peace. 'Ask Harriet to put the kettle on, please Gigi? I'll follow you once I've totaled this row of figures.'

In the mounting yard, Bjorn's breakers and stockmen either squatted under or leant against the trunks of ghost-gums to drink their tea. All eyes focused in the direction of the property next door. Abergeldie's long drive stretched a quarter of a mile from the road, on Jalna's left. A huge patch of land, cleared around the house was fenced off and hidden by groves of poplars that shaded the homestead and allowed its occupants privacy. The property's home or lower paddocks, an immense area, were ideal for ewes and their lambs to graze on rich pastoral grasses. The top paddocks meant their rams were protected from frosts and strong gale-force winds that swept in across the southern plains. Abergeldie ran thirty sheep to an acre, equal to the Svensson's spread.

Most of the shearers were drifters seeking better pay and refuge from severe summer heat and strong westerly winds while working though their circuit. The majority worked in conjunction with Abergeldie, the massive property Bjorn's men were now appraising. The homestead and its huge expanse of fertile land were on the market. Abergeldie was located eleven miles from Yass and fifty miles in a crossbow-line from Canberra. Bjorn reckoned this distance was a mere hop for a big red kangaroo, compared with the entire Australian landmass. The small homely township was situated in the south-western quadrant of New South Wales. From Yass it took two and a half-hours to drive to outer suburbia of the capital, or an hour and a half's drive east to Goulburn.

After the women returned indoors, Bjorn's men discussed Abergeldie. The property for sale adjoined Jalna's southern flank and right boundary.

'Ian, what's your honest opinion of Abergeldie? I reckon it's a spread and a half. Well worth the money.' Envy elevated Bjorn's voice. An ideal property which he and Kirsten wanted to purchase. They would run it in conjunction with Jalna to produce different breeds of sheep, apart from Merino. However, it wasn't to be. Their cash flow stretched just far enough to keep Jalna viable. At present their finances were tied up with running the property, horse stud and racing stables as well as other land holdings. Lacking surplus funds to compete with his competitors, it prevented Bjorn from bidding for Abergeldie.

Bjorn's four thoroughbreds were stabled at his trainer, Buck Master's stud all through the summer months. He lived at the opposite end of town. Buck was prepared to risk backing the Ross family financially, to relieve Svensson the pressure of being a liability to Ian. This was disputed by Bjorn, who informed Masters that their funds were secure.

After smoko, he conveyed this news to Ian Ross who didn't bother to respond. His eyes hadn't deflected off the staunch figure standing alongside him, also leaning on the fence. Bjorn preferred to ignore his arrogance.

Every now and then Bjorn stubbed his well-worn blucher against the upright timber. His effort to dislodge the small bit of gravel digging into his big toe failed, much to his disgust. With that boot resting on a wire strand near the wooden post, and his other one firmly planted in the dust, Ian scowled at him.

'For God's sake remove that damn boot if it's worrying you,' he snarled as his eyed travelled over Bjorn's lean figure. 'You asked a moment ago my opinion of Abergeldie. I've given it a lot of thought actually, watching you bash hell out of the fence. If you're trying to knock the bloody post down, you're doing a damn good job. Those frigging stays are loose.' Casting a frown in his boss's direction, Ian could feel his anger climbing to its pinnacle, until he sarcastically bellowed, 'What you're doing is annoying. I'm not standing here, if you continue with this damn stupidity.'

The day hadn't gone well for the under-foreman. Constant bickering between him and Bjorn had inflamed both their tempers. A cloud of grit stirred up by the shuffling boots engulfed them both. With dust swirling around his face, Ian tried to protect his eyes with a tattered Akubra another bloke had dropped. A thundering headache felt as if it would split his skull apart. The stinging wind from a willy-willy bit deep into their sunburned flesh. Fine particles of powdered grit shrouded the men's bodies and sweat mingled with sludge trickled down their backs.

'Must you? There's enough muck flying about now without you stirring up more.' Ian coughed, furiously waving the hat. The bitter taste of airborne earth was repulsive. Thick in patches, it invaded his lips and nostrils. Smothered in filth he spluttered, 'Shit, it's bad enough standing in this heat without you spreading more blasted muck. You're an inconsiderate bastard, Svensson.'

'Give it a break mate,' he growled. Bjorn objected to this employee's nasty inference and removed the offending stone from his boot. Clutching the top strand of wire, he shook his foot and it freed the pebble. With his blucher back on, he wriggled its leather cuff above his heel until it felt comfortable. Stamping that foot, he observed the wrangler breaking-in a gelding then cursed having missed watching the previous effort.

The young breaker was plying his trade with Bjorn's prize purchase. Annoyed he frowned as Ian, in a huff, headed towards the bunkhouses.

'Ha!' Bjorn scoffed to Lanky Hancock, who'd taken Ian's place at the fence. 'I bet that smug bastard's gone to wash his gob, before the meal gong shatters our peace. Trust him to get in first. There won't be a bar of soap left when he's finished scrubbing his filthy paws. As long as he doesn't get stuck into his wife, I couldn't give a shit what he does.'

'Boss, I don't know how you can stand that bastard. I tolerate him for her sake. I'll kill him if he bashes her again.'

'Gigi's taken about as much as she can stand from Ian Ross. Have you heard him swear in a foreign language? Why I asked, the bastard muttered something the other day about Berlin. Drunk as a skunk, I missed the gist of what he meant.'

'No. I heard him swearing in what I took to be Austrian. I wouldn't put it past the bastard. If I hear anything weird, I'll let ya know boss. Then we'll find out once and for all where the mongrel originated. He reckons he's Irish.'

'Not with the Scottish surname of Ross.' Born smiled. *I might sound Gigi out. Ian's a compulsive liar. Why he takes his agro out on her mystifies us. My wife might have a clue of his real identity. I'll tackle the problem head on tonight in bed.*

Before heading back indoors to enjoy lunch, made by their relieving cook, Harriet Graham, he spoke to his men near the breaking yard. There Bjorn watched his latest acquisition to his stables being broken in, before a trainee breaker, Corey Sidle walked the yearling back to his stall.

Delta Boy showed promise of being a terrific stayer on a full course. This afternoon Buck Masters had brought the colt down for Bjorn to trial him on the flat track behind Jalna's stables. Both he and Lanky were anxious to see how well the sorrel handled on his five hundred metre racetrack.

A long search of the countryside around Yass had proved beneficial when he discovered this talented young boxer with terrific prospects. This time BJ had come up trumps. Satisfied with Delta Boy's performance, he ordered a stable hand to take the colt back to his stables. Corey had gone with his boxing trainer, back to Yass. One of the young muckers would give Delta Boy a good rub down and sand bath, before Buck returned to collect him.

After their half hour lunch break BJ's crew sweated on watching the next horse being broken in by another a novice breaker.

'If you don't want to stand here with us, you can leave, if you want to Ian,' he declared to the employee standing by his side.

An attraction for this horse and relieving wrangler had inspired Ross. 'Not on your life. The kid doing the next stint is new to breaking.' Observing the horses perform took Ian back to his younger years in Germany when he trained at the Calvary School in Hanover. Yet he daren't disclose his Nazi heritage or ever serving as a captain in the Reich to anyone, especially not the man standing beside him. 'Besides, I want to observe the way you handle both him and the horse. It should be interesting.'

'What, me, the supervision or both? Do you mean about this procedure, or my way of managing the lad? Make yourself clear. I won't tolerate a rebuff of my methods of breaking-in from anyone, especially not from you.' BJ made it clear that he wouldn't accept criticism from his men while any procedure was in progress.

Ian responded to the chiding with a sleazy look at Bjorn. 'This exercise of breaking I meant. You damn well knew that. It looks like each breaker has his own style and method. I'm interested to see how this youngster handles your horses. That's all, mate.' The inference of mate in his voice stank of ridicule.

On the verge of telling Ian Ross, his under-foreman to shut his mouth, Bjorn remained silent as together they watched the trainee breaker working this horse. He didn't wish to start another argument like the one he'd experienced a week earlier. At the time Bjorn had pulled Ian to account over his belligerent attitude to his wife and their staff. In retaliation and inebriated Ian had physically abused Gigi, by twisting her arm until the pressure applied had dislocated her right shoulder.

Today during their afternoon break Bjorn again warned him.

'While you're working on my property, I expect you to show me and these men some respect. That includes my wife and your own. I will not tolerate your abuse, Ian.'

Excusing himself, Bjorn stormed off in a temper. *Belligerence is the main reason I encouraged that ignorant bastard to buy his own stud. If he's successful in purchasing Abergeldie and if the sale goes through, it'll keep his mind occupied and those damn hands busy. I know he's keen to take over and run the property without interference. He's always criticising me and my methods in managing Jalna's mob. Well, he's going to find the property next door a real challenge.*

This entire business was instigated by the Svenssons, so they could observe Gigi and Kendall at close range, when the lad came home from school in two weeks. In their front hall Bjorn approached his mother.

'Gigi, you and Kendall are welcome to stay here on Jalna until your husband has Abergeldie working to its maximum capacity.'

Having overheard him, Ian's reply came as a grunt. She didn't decline Bjorn's offer and waited until her husband stomped off.

'I'll be grateful if Kendall can stay here with you and Kirsten. It'll prevent arguments between him and Ian. As you know Bjorn, they antagonise one another for the simplest of reasons. It will save me a lot of worry.'

Mrs Graham appeared as if from nowhere and Bjorn signalled her to wait. 'Harriet, is your cousin Helen well enough, to resume cooking for us here on Jalna? I'm not trying to get rid of you, far from it. You've done a terrific for my crew and us.'

'Yes, Mr Svensson she wants to return at the end of this week. If you don't mind me saying this, if Mr Ross and Gigi do buy Abergeldie, I want to be their housekeep and cook. Then I can keep an eye on Gigi. I don't trust Ian. Heed my warning, sir. Make sure he doesn't come anywhere near the house if he's been drinking. Keep a firm watch on him.'

'What made you say that, Harriet? Have you seen him hanging around here when I and my men are up working in the top paddocks? Or mending the lower fences?'

'You know I'd never be a tattler Mr S. It's just that I've grown quite fond of your wife and Gigi. They're very special ladies to me. I would like to leave here when Gigi moves over to Abergeldie and so does my daughter. Rose can work in the kitchen with me until they find other, *not so young* girls to do their housework. There are a lot of beds and rooms to keep tidy over in that huge homestead. A couple of my friends, their adult daughters need work, and I'll recommend them to Gigi. Not Mr Ross. He's crude. Every time his mouth opens out drivels horrible swear words.'

'Harriet, I'll need to ask Gigi first. I'm sure she'd love to have young trustworthy girls doing on the duties a household the size of Abergeldie will need. Most of the present staff have either left, they or intend leaving next week.'

Worried about her health Bjorn spoke to his wife before tea. 'I'm sure Gigi will be abused again the instant Kendall steps in that door. He and his father argue and dislike one each immensely. Kendall's passive whereas Ian flies off the handle constantly. The present owner of Abergeldie wants a quick sale. Ian better decide soon, if he wants to buy the house and property. Kirsty, he's a close-mouthed bugger regarding his private affairs. He resents me telling him anything. This morning he warned me to mind my own damn business. Arrangements are in place to buy it, or so he reckons.'

Bjorn fumed while putting his boots back on before stepping into the back yard. *I know more than the bastard's letting on. Why the hell can't he be honest and tell us the truth?*

Standing in shade to observe how the breaker was handling the current horse Ian checked his watch. *If I'm late, that fart-arced Beaumont may foreclose on the deal.*

Bjorn looked across at his stern features and smirked. 'What time's your appointment in town with the solicitor?' Patiently he awaited a reply which he expected to be rude or hostile.

'Ah, wouldn't you like to know? I have a minute or two before I leave here. Don't ask again. It's none of your business, Svensson.' In a shitty mood Ian sniggered, 'I'll go when I'm ready, with or without your permission.' This caustic response reminded Bjorn of a snake ready to strike. He again cautioned Ian not to walk in the house unannounced.

'Now answer my question. What time do you and Gigi have to finalise the deal on Abergeldie?' He had a fair idea it could be five.

'Five-thirty, if you must know. Curley Beaumont won't be in Yass until then. I bet those damn women aren't home yet. I need my car.' Ian's temper sharply changed for the worse, if possible. 'My wife had better not have wasted my cash on clothes. She'll feel the rough end of my boot if she's used a cheque from my account.' This rebuttal of his wife held a narcissistic ring and was unwarranted. Angrily Ian stroked his grey beard. It had grown from a lean crop of wispy hair to a neatly trimmed goatee.

'I'm eager to see you treat your wife with respect. Leave her alone, Ian. Haven't you done enough damage?' Bjorn clammed up. In a belligerent mood he might retaliate and take his animosity out on Gigi on her and Kirsten's arrival home.

'Go on, I dare you to call me mate,' Ian sniggered and spat through clenched teeth. 'Come on sport, I'm itching for a fight. Or are you a gutless twit?' Ian hissed and his arrogant pose caused Bjorn to think of an ally cat; claws drawn, ready to pounce.

'Don't forget you're still on our payroll and on my property. Remember that you smart-arsed nerd. I've had a gutful of you sniping at me, my wife and your wife.' The anger in Bjorn's raised voice struck home. Frustrated, he removed his hat, scrubbed his partly-bald head and then reseated his battered slouchy. It had seen better days.

With a discourteous glower Ian snarled. 'It has nothing to do with you, how I treat my wife. I object strongly to your derogative tone. You really are a bastard, Svensson.'

He ignored Ian's antagonism, and dismissed his rudeness as a product of ill-breeding. Ian's spontaneous display of temper he deplored. Having exploded himself, he felt great.

Aware he could block the sale of Abergeldie riled Ian. Bjorn and the owner were the best of mates. With several local farmers interested in the property, Ian knew it would be foolish to inflame this situation by the continuance of their vendetta.

Another time I can hobble that smart bastard's tongue. As of now I prefer to let his blatant remarks pass. There's positive news in the pipeline that, in time, will test his patience. That brat of mine better not give me cheek this weekend. I'm not travelling to Goulburn. His mother, or this arrogant prig or his wife can fetch him. Bugger standing here to converse with this idiot. I can do that at my leisure.

Distracted, he focused on the next door property for sale. *They haven't a clue that I've taken a photo of their record ledgers, before that snoop, Harriet Graham caught me near Kirsten's bathroom. Why didn't I rape the bitch when I had the chance? She taunted me by swaying those naked hips. She almost dropped the towel when she realised I was standing there. Her figure resembles Venus de Milo. Not bad for a woman in her mid-forties.* Ian's mind diverted to the missed opportunity of photographing her naked body. *A snap of her curvaceous breasts would be an ideal tool to blackmail her idiot husband after I own Abergeldie. No bastard will underwrite my bid for the property. It's almost sealed and delivered. Once I sign the final document this afternoon, Svensson won't be game to poke his ugly mooch inside the homestead's door. I'd kill him first.*

Ian knew the Svensson family had contacts all throughout the southern districts. If things went haywire, Bjorn could call on trained men for the positions of manager and head foreman if necessary and at a moment's notice. Earlier in the week he promised to line up a few outstanding blokes within the field of expertise a sheep property the size of Abergeldie required. *I know Lanky, Jalna's ringer and top-gun confirmed in my hearing that he'd continue working both properties. Perhaps if I accept their advice, I'll learn the fine points of wrangling from Svensson, even though he constantly infuriates me.*

Procuring reliable household staff would be a huge problem. Without Kirsten's help to recruit extra female staff, Ian knew they would flounder. Women ran the properties on their stomach and were important. A condition of sale was for the purchaser to keep all the staff on Abergeldie. Ian Ross, being the preferred buyer of Abergeldie and its holdings, he was obligated to comply with this and a long list of conditions.

As Ian strode past her office window Kirsten, an observant woman, was concerned: *I'm positive he comes from a military family. My father, a colonel in the last war, analysed his problems and mulled over the consequences before he tackled them. Ian's methods of solving difficulties are identical to dad's even when he's inebriated. If something serious occurs, it sobers him instantly and he copes and solves that problem.*

'What's on the agenda this afternoon, boss. More horses ta broken in?'

'Sure is, Lanky. Tucker's ready boys. Be out here at one, we can't afford to waste time this arvo. There'll be ten minutes smoko, then back to work.'

14

1.30pm in Jalna's mounting yard: Bjorn stood in the shade to caution the eighteen-year-old boxer to be wary of the horse.

'Ponch, don't let get him get the better of you, or you'll regret it lad.' Observing how he calmed the bay he held his breath. 'Bring him around this way,' a hand curved in the direction, 'good, now release him. Careful, watch that left hind hoof. Sun Bronze will clobber you, given half a chance. He's rearing to go.'

The horse was feisty and wanted his own way. BJ and this handler had other ideas. Neither of them believed in being cruel to the animals in training. By taking control at a specific point of time it eliminated the need for a whip or brute force. Ian had a different opinion on breaking-in horses or training dogs.

Corey Sidle, the prize fighter whom everyone called Ponch, stood his ground. He took notice of his boss' advice and the correct way to break a horse. Bjorn knew with this knowledge under his belt, the lad might make a terrific cropper in time.

Sun Bronze glared at his trainer, then snorted and dug his hooves in the powdered earth, sending dust in all directions. With their backs facing the compound and standing fifteen paces away the stockmen coughed and spluttered when swamped with a flurry of grit. Bjorn turned his head away to spit out a voluminous stream of red grit.

On recovering enough to watch the procedure in progress he called in a moderate tone to prevent the horse from spooking. 'Keep clear of that hind hoof Ponch, or your knee will be the recipient of its back swing.'

Wiping his lips free of dirt on the bandana around his neck, Bjorn tucked the sweaty cloth back in the hip pocket of his saturated jodhpurs.

When able to speak, he nodded for his head foreman to move in closer. Bernie Staples, a big-bronzed Aussie removed his glasses, wiping them on the tail of his kaki shirt. In gay abandon, he flipped the weathered slouchy against the knee of his patched trackpants. This dislodged yukky sludge. Shoving his hat on, a tap on the crown settled it on a well cropped scalp.

'Close in slowly I said. Listen, Ponch. Now ease the rope off. Plenty of rope, that's right.' BJ's soft tone settled Sun Bronze. Given leeway he seemed calmer. 'Now back off as if you're not interested. Then without letting him know, keep an eye on his movements.' He followed each individual stance, of both the horse and his wrangler.

Jalna's foreman settled his boot on the bottom rail and remained silent while Bjorn was giving the lad instructions. 'Hey boss, how did that smart-arsed big-wig go? Do ya honestly think he'll make a go at working his own plot over there?'

'Ah, you mean Ian. Good tag for an obstinate prick. He caught on to some things. Mind you Bernie, only the ones he wanted to. The financial side of their business Gigi will be handling. I needn't tell you why.' Bjorn brushed a bunch of flies from his eyes. A march fly had stung his little finger. Irritated, he kept scratching the inflamed welt until the skin was red raw.

'Won't bugger-lugs buck, with his wife controlling the books? I wouldn't like to be around if she makes a mistake. Knowing the bastard, he'd chop her fingers off.'

'Hardly! He won't be game to abuse her again. He's aware I'll pull the plug on being their guarantor for Abergeldie. There's today's payment and the final papers to be signed before the sale goes through and settlement. And he knows it. I've drummed that fact in his dumb skull enough.' Pausing, BJ reflected on Ian's sneakiness regarding the purchase. 'Bernie, he'd be a fool to go against me. The man's laying out a lot of cash on this investment. If he isn't the successful bidder for that property, I'll soon send him packing. Then I'll let his wife work on Jalna. Kirsty and I have offered to let Kendall stay here every semester holidays. Or until he's old enough to make his own decisions.'

This blasted sting is driving me crazy. His fingers tore at the huge blood-red welt on weathered skin. The itch was unbearable, even for a bushy like BJ. Retrieving his shirt from under the tree he shook it well before sticking it on his sun-hot shoulders. The flesh on his limbs had a slight glow of ochre. Deep olive-skinned, BJ could be mistaken as a distant relative of their aboriginal backtracker Jamie Ludawalla.

'One never knows in the wool growing business when help is needed from a friend's pocket. More than once I've needed my uncle's dough and his wisdom to get me through a scrape. Bernie, we'd be broke without Sam and Edna's help. They've been our backers since day dot. Well, until our finances were stable enough to repay our bank.'

Reefing off the Akubra he fanned his face. It cleared the air of pests and insects. 'Hey Bern, aren't bluebottle flies a sign of rain coming in? We might get a deluge before nightfall. That blackish-green anvil rolling in over the southern escarpment looks mighty ominous.'

The men raised their eyes westward. Bjorn replaced his hat and tapped its crown to secure it from the rising wind.

'You know BJ, when that bastard flies off the handle his anger is fierce enough to blow the top of a chimney. One day soon a frustrated husband will shoot Ian for being cruel to his wife. I don't trust Ian Ross. He'd dip his wick in anything with or without pants. He flirts with your housemaids here on Jalna. One gander at his battered wife is enough for me,' Bernie confided, swishing flies away that were pestering his eyes.

'Bernie, I shouldn't say this,' Bjorn glanced around to see if his men were within a cooee's call. 'The quicker a rope tightens around his neck, the better for everyone.'

'Why do you keep helping the idiot? I wouldn't give him a second of my time.' Sneezing, Bernie slammed his left boot on the fence rail. Dust went flying everywhere. But it dispersed quickly.

Together they angled back to the house. On pay day the men finished early. Bjorn continued the conversation on the homestead steps. 'For two obvious reasons, I shouldn't have to name. We deeply care for Gigi and Kendall. They're family to us. That dear boy...' Bjorn covered his nose with a gritty hand. The earthy smell of powdered earth affected his olfactory senses, and caused a whopping sneeze. Fragmented bits of grit flew into his eyes. He relieved the itchiness by cupping them in cool tap water.

'Bless yer mate,' laughingly affirmed his foreman. 'I actually saw the cruel mongrel strike Kendall with the buckle end of his belt. Ian better watch his backside. The lad almost towers over his old man. How old is the boy now BJ?'

'Fifteen in December, and he's growing taller by the day. He won't tolerate Ian's drunken tantrums much longer. Unfortunately, the lad has to knuckle to his old man's demands. Am I sweating on the day ...' Bjorn hesitated to seek an answer to a topic which had eluded him. 'I meant to ask you, Bern. How's my new colt? Broken him in yet?'

Bernie nodded as if to say, *you better believe it.* He knew the bay would be Bjorn's best stayer. Then inquisitiveness got the better of him. 'You going to track him?'

'Race him, not on your life,' BJ acknowledged on their way back to the house. 'He's Kendall's horse. That's his decision. The lad can please himself what to name the colt, after he's gelded.'

Bernie Staples nudged his sleeve. 'Shush! There's his old fart now. He just turned that far corner. I think he must be going ta meet his missus. Na, he's looking this way. Shut it BJ, unless you want him to put the kybosh on your plans for the boy.'

Bjorn smirked and spoke with a mediocre lisp. 'If Ian spoils our surprise, I'll use his guts for a shanghai then present it to his wife.'

They laughed as Jalna's manager smirked. 'Somehow boss, I can't see her using a catapult. She thinks the sun shines out of his arse. Why, I can't imagine.' Bernie flicked a spot of rain off a bald scalp as a fly buzzed on the tip of his nose. 'Good,' he said with an Aussie flagging gesture. 'Looks as if we'll get the predicted downpour soon.

'Yeah. I'm grateful that you yanked Ian away after their last debacle. Bernie, I had no option but to intervene.' Bjorn lowered his voice. 'Otherwise, he'd have killed his wife.' Interrupted by Ian, he refrained from saying another word.

'We're away now. Gigi's in my car. Your wife wants you.' No address or courtesy was forthcoming to his boss. The men never expected one, just flexed their eyebrows.

'Tell Gigi the best of whatever. We're thinking of her,' Bjorn casually affirmed, sporting a scowl 'Oh you too, of course. I won't come out mate.' Any form of blessing or courtesy would be wasted on Ian Ross.

The cryptic remark sounded so unlike Svensson. Usually he smoothed things over and included Ian in the conversation. Bernie disapproved of his supercilious smirk and looked him up and down. Ian was dressed like a toff. His beard was trimmed. Watching him strut to the car he scoffed. 'That was a bit blatant, boss. I know the rotter needed pulling down a peg. Ian's an arrogant swine. How his wife puts up with his rotten tempers, stumps me.'

Having seen the state of her after her recent *unfortunate accident*, Jalna's foreman had to agree with his boss. And knew Bjorn was justified in condemning Ian.

'Hey, it'll be interesting if he buys Abergeldie. Ian's solicitor's a shifty bastard. He reckons Jaffa Thorpe needed to sell, ta keep his finances on track. Ha, we both know the real reason.'

Bjorn tilted his head.

'I tried to advise him about Murdock's devious practices. He's a shyster and a drifter. I'm damn sure he's not licensed or registered. As usual Ian refused to listen. After they settle in, I'll persuade Gigi to transfer her personal documents to our solicitor. Harry Ticehurst, my financier is a bonza bloke. He handles all of Jalna's legal business. My uncle, Sam Houlton put us onto him.'

Finished for today, the young breaker walked his latest acquisition to the holding yard to groom the stallion before Buck Masters returned to collect him. One of the young muckers on Jalna's payroll had already swept and hosed the horse float.

Bernie stood beside his boss at the homestead's rear steps. And as Ponch stopped Bjorn spoke to him. 'Tether the horse to that macadamia tree and come here lad.'

Securing the reins to a sturdy bough and in his manly stride Corey sauntered over to both the foreman and his boss. Bjorn's warm smile encouraged him to think positive for tonight's bout.

'Stand here with us for a tick, Ponch. You excelled yourself this afternoon and I'm proud of the way you handled all the horses. There's a bonus in your pay packet.' Bjorn tapped the young boxer's shoulder.

His face beamed and moving up a step he leant on the rail their bottoms rested against. Sweat trickling down his back, he wiped his brow free of moisture.

'Phew what a corker of a day, working under a blazing sun. Hope tomorra will be better.'

'Cory, you and Mark Sarrison are good mates. Keep an eye on him. Being one of our new handlers, he works a full shift without a growl.

He can walk Sun Bronze back to the stables.' Bjorn motioned for him to return the horse to his stall.

'Enjoy your weekend off, Ponch. You've earned it.'

'Yeah, and I'll shout you a pint later Boss … and you too Bern. Better get going or I'll be late for training. Then I won't be in good nick to tackle my opponent tonight.'

The lad didn't wait for a reply. He slung his butt on the horse as Jamie, the intuitive native backtracker trotted his roan out of the stables. Bjorn saluted the terrible two as they disappeared up the tradesmen's road. Together they rode to greet a brilliant sunset.

'Bernie, I think Corey's got the makings of a professional breaker. I hope he sticks to it, and not boxing. He's a keen man in that field. In my honest opinion, he'll go a lot further with wrangling. Has the top shearer in his sights, so he reckons?'

'Yeah, Boss. With his skill, he has the guts to beat our top-gun. Who tagged him Ponch? Corey Sidle's a clean fighter, he wouldn't take a bribe. He and my teenagers are the best of mates. They seldom argue and get on famously.'

'His peers did.' Bjorn thumped the rail with a closed fist. 'Struth, he's gone without collecting his pay. Bernie, you've a ringside seat tonight. I'll fetch his packet from the office. Will you give it to him before he enters the ring?'

'Okay. I want to give the kid a surprise. Ponch hasn't a clue I'll be ringside tonight. He reckons we've got a terrific crew here on Jalna. And I agree. There's only one fly in the ointment, the idiot who digs his heels in if asked to do some hard work, aye boss. You know who that bastard is BJ.' They winked in tandem.

Bjorn's glower tagged his thoughts. *Yeah, he delights in trying to sabotage our progress and ridicules my authority. With a bit of luck and a hundred thousand well-spent quid, Ian Ross will soon be a bitter memory to our crew on Jalna.*

Bernie's churlish grin implied what Bjorn had presumed. They watched a strapper walk Sun Bronze to the mounting yard to give him a good rub down and roll in the sand pit. Securing the stable doors, the boys then headed to their bunkhouse for well-deserved showers.

Before stepping over the homestead's threshold Bjorn removed his bluchers, giving them a tap on the iron scraper anchored in cement.

When satisfied their soles were free of muck he put them just inside the wire door. To keep the boots spider-free he tucked a sock over each rim. This also prevented snakes and vermin from bedding in them.

Tossing his hat on the table, he stopped by Kirsten's office to see what she required. Busy writing, she looked up, nodded, then continued talking to someone on the phone. Without a word BJ returned to the kitchen.

'Well, that's work finished for today. Helen, it's good to have you home with us. Your cousin Harriett had an appointment in town this morning. She home yet? I'm thirstier than a paddock in drought. Been a hellishly hot afternoon and everyone's flat-strapped as we say and you, being a Pom know our Aussie expression well.'

The housekeeper laughed. She knew her boss was joking and he felt this dry heat and needed a cool beer and she put four stubbies on the table. 'Hey, they're not all for you, Mr Svensson.'

'I know that, Lanky will be here in a tick. It wouldn't take a bush lawyer to see how we're all melting in this heat. Fetch a few more bottles of light ale from the first cool room, please Helen.'

Sweat gushed down his back. It squelched through his shirt as he lent against the chair. Bjorn mopped his brow to stop the salt stinging his eyes. Then relaxed and expelled long drawn-out sighs through parched lips, bronzed by the hot summer wind.

'What a day it's been. Ian Ross is a stubborn man. Did he say what time he and Gigi would be home? If not, put their meals aside please.'

Helen Raddick couldn't answer, not with a plum in her mouth and they'd grown some beauties this year. The seed catapulted from her warm lips down to the bin.

'My mother taught me not to speak with a mouthful. And don't tell me to fill it up then speak, like you usually do.' Bjorn, Helen and her cousin, Harriet were the best of mates. Jalna's male staff and the local shearers were envious of their friendly rapport.

'I'll let it pass this once. I forgot, Bernie will also be in shortly. Here,' he pushed two bottles across the table, 'keep those on ice until he and Lanky land in. There's nothing worse than a warm beer.' A finger wagged as Helen grinned. 'I know where you hail from it's the usual thing. That's for living in England most of your life.'

His taunting went unnoticed. She put the treacle and date pudding in the fuel oven and lowered its thermostat ten degrees.

'Gee, will I be glad when our manager returns tomorrow.' Bjorn breathed heavily and took a sip of the ice-cold brew. 'Bill needed this break. A whole week off.'

'Mr Kearny, will he be here for dinner? If so, I better put more vegies in with the lamb roast. I'll bake another rice pudding. There's plenty of fresh bread in that chipped crock. If you don't mind, I'll get a new one in Yass tomorrow?'

'That's okay, Helen. It looks like Doc Jarvis will be here tonight. Better than him going home to an empty house, or working all night in his surgery. He's running low on the sterile syringes. I broke one, don't tell my wife. Kirsty growls over my clumsiness.'

The casual courtesy Bjorn showed to Helen, including Jalna's guest, Gigi was typical of his nature. If he couldn't be kind to someone, he wouldn't be nasty or harm them. He considered his staff the best this side of the Great Dividing Range. A sincere friend to those in need, he'd walk miles to help a fellow bushy or strangers in trouble. The Svenssons valued Helen as a family member. They appreciated her and Jasmine's artistic skills. Treasured commodities, this daughter and mother duo were their mainstay if they were in trouble.

Having witnessed Ian Ross verbally abusing his wife that evening, neither Helen nor Bjorn appreciated his rebuttal. He disapproved of the staff being rudely spoken to by any outsider. And Mr Ross was no exception.

Kirsty mistrusted Ian for his habit of snooping, after seeing her naked. He resented the chiding by her angry husband for a minor indiscretion. BJ suspected this was the reason for his uncouth grouchiness until now, with sunset on the wane.

Since the lunch break Ian's mood had deteriorated dramatically until several of the roustabouts had threatened him. Burly men in their thirties, the strength in their biceps was powerful enough to kill a man with a mediocre clout. Ian took the warnings as he accepted everything, as a grain of salt in a turbulent sea of lies. In return he threatened to harm his wife, if the men didn't mind their own business and keep off his patch.

Helen received a healthy bonus every time Ian demanded her to do his laundry. But only if his wife was ill. Indignantly he objected to his pay being docked after indirectly causing a fight in the woolshed breezeway. Gigi became the recipient of that money. It amounted to thirty-five quid.

Having his meals restricted didn't faze Ian Ross. He preferred to feast on whiskey. The drunkard reckoned a good drop was better than lack of privileges and food.

In one of their recent arguments BJ was tempted to tip his cheap grog in the sink and substitute it with cold tea. After a night of binge drinking, Ian was so inebriated that he couldn't distinguish the difference between cold tea and bilge-water. Should he discover an empty whiskey, or gin bottle in his hoard it would incite World War Three. In his wisdom, Bjorn abandoned the idea.

Dinner over, everyone relaxed in the lounge room. In comfort they listened to the radio playing popular music, until disturbed by car lights flooding Jalna's drive.

In her quiet and sedate manner, Kirsty moved to the front wire-screen door.

'I think they're home now Bjorn. No, it's Gigi on her own. Where's Ian?'

'Sorry I'm late Kirsty. We were having coffee in Stuart's Continental Café, and Ian objected to its strength. He argued with me and then stormed out. And left us to pay the bill. His solicitor bought me home. Is Ian here, Kirsty?'

'No. It didn't look like his car.' Arms outstretched she welcomed their guest home.

Gigi took her time hobbling up the front steps then embraced her best friends. 'Guess what! You're looking at one proud owner of Abergeldie. Isn't it terrific news?' Enraptured, she lost her balance and instead of falling down the steps she fell into BJ's arms. 'Let's wet the baby's head.' She struggled to pull a corked bottle of Chablis from her handbag. Giggling hysterically, she held it aloft. 'Pour me a drink in your best champagne goblet, Bjorn boy,' Gigi slurred the words. 'Is that red-eyed monster in this house? If he is, don't let him come near me. Look,' her voice rose, 'there's spiders swarming all over the curtains and walls,' then it lowered to a whisper. 'I can see snakes on your collar, BJ.'

'She's hallucinating, Bjorn,' gasped Jarvis. 'Whatever drug she's swallowed, her breath indicates it's not cyanide. Once she's settled in bed, I'll take a swab of her sputum. If we support her arms, Gigi will be able to walk to the first bedroom…'

'Excuse me interrupting, Stephen. The sheets on that bed were changed this morning. They're turned down. Put her on the bed while

I get a bowl of warm water to bathe those bruised arms. Be careful of her knees Bjorn, they're swollen. So are her ankles. Going by the congealed blood on her skirt, she's either fallen in town, or a certain drunk we both know has lashed out with his closed fist again. Her breath smells vile.'

'Kirsty, call Lanky or Berny Staples. They're fixing that wonky tap in the laundry.'

'Bjorn, are you drunk?'

'No, I am not Kirsten. Haven't touched a drop since Christmas. Well, I drank a glass of vintage wine your equestrian tutor bought last night, at dinner.'

She filled the five litre kettle with filtered water. With three cement tanks running dry, every drop of this precious commodity must be boiled. Her foot pushed the door open just enough to enter the bedroom. 'Quick, I'll drop this jug if you don't hurry Bjorn. It's heavy. Grab the porcelain bowl from under my arm and don't drop it. They're heirlooms.'

'They'll be safe on this table by the bed. You did boil the water, twice? Doc Jarvis's orders. Not mine.' Stephen looked at the sterile swab in his hand and grinned.

'Yes, three times with cold water. I filled the urn and left it cooling on the kitchen sink, Born. For being smart, you can carry it out to the closest cool room.'

'Okay, I will later. Has Lanky or Bern landed in yet, Kirsty?'

'Arthur's in the hall. Stephen, do you want me to give him the swab in your hand?'

'No, Kirsten. I'll use the hall phone to buzz the surgery. Either my locum or a nurse should still be there.'

'It's fine by us, Doc. We're expecting a call from Kay Brown. She offered to collect Kendall, with her son in the morning from Goulburn.'

On his return Jarvis passed the tagged swab to Lanky Hancock. 'My receptionist will still be in the surgery. She'll give you a package. Yay size.' Jarvis spanned his hand. 'It's breakable. I can't leave here, not while Mrs Ross is so ill. Do you mind, Arthur?'

'Of course not, Doc. My pleasure to do it. You're always chasing everyone's tail here on Jalna. Ya parcel will be safe in me saddlebag. Tell the boss, I'll be back before he locks the cool rooms, or hits the sack. What time are you turning in?'

'The moment I've given Mrs Ross this injection. Arthur, I'm trusting you to deliver this parcel unbroken and in good nick.'

'It'll be as safe as a bank, Doc.' A curt wave and Lanky Hancock hurried out to the back yard where he'd tethered his horse. On his saddle he waved.

'We'll be off soon, Kirsten. My son's car is pulling into your drive. Harriet and Rose are reading in their rooms. If you need anything they'll fetch it for you. Oh, how's Gigi? Her face was drained of colour and she looked really ill this afternoon.'

'Jarvis has just sedated her. Thanks for keeping his meal hot in the warming oven. I'll get it. And I won't come out Helen. Give Paul my regards and enjoy the night. *Blossoms in the Dust* is a beautiful film. You'll need a few hankies, it's a tearjerker.'

'We will. It's a special screening at the schoolhouse. We should be home around nine, or a little after. Unless Paul wants to stop somewhere for coffee, bye dear. Have I forgotten the tickets? No, they're here in my purse. See you in the morning.'

Collecting her coat, gloves and handbag off the sofa, Helen Raddick eased the back door to a close and cut through the cool room breezeway. Until Ian barred her way. The angry look in his eyes frightened her.

'Where's my wife? I bet she's in there with them. Move out of my way Helen, or I'll clobber you with this book …'

A large hand reached out of the shadows and gripped his neck. 'Who might you be, abusing my mother?' Under the coolroom light, the intruder recognised Ian. 'I know of your reputation Ross. You're the rotter who's been traumatising young women in town.'

A flash of moonlight reflected on the badge under his lapel. Ian gasped. *Trust me to threaten a copper's mum. Struth, if he doesn't release my shirt, the bastard will choke me.*

'Remove your hand, or I'll knee you in the groin. I wanted to know if my wife was in the house. I didn't threaten Helen. Can't a man ask a question without being manhandled by you?'

'I caught you threatening my mother. If you ever try that again I'll maim, not kill you. In my field of work, the term is self-preservation. The law and thugs like you call it self-defense. Either way it's negative. Now get out of my sight, or I *will* throttle you with my bare hands.

You're despicable.' He slammed Ian's skull hard against the brick wall. It rebounded, but the wall didn't.

With a throbbing head, he supported buckling knees in both hands. Ian writhed in agony as DI Raddick escorted his mother the car. Travelling up Jalna's drive, Paul acknowledged a dark-clad rider as his horse loped towards the house. 'What's Lanky Hancock doing out ...'

'He ran an errand for Doctor Jarvis.'

'Ran an errand on a damn horse? Mum!'

'You knew what I meant, Paul. A slip of the tongue. Let's enjoy tonight, not spoil it by arguing over trivial incidents. That's one of Ian's perks.'

The school hall was crowded with film-buffs. Tonight's proceeds boosted the Muscular Dystrophy Fund. It enabled them to purchase equipment in Canberra for disabled patients in Yass and the surrounding districts.

Doctor Jarvis, one of the fund's original patrons had given his apology by phone. "An overburden of work, due to the raging influenza epidemic will prevent me from attending your preview." Little did he know then, that tonight his time would be spent trying to save an unconscious patient from choking on her own sputum.

1.30 am. The pen dropped from cramped fingers. The nib piercing her chart. Royal-blue ink splattered the drowsy physician's spotless shirt, stained with perspiration and sprinkled with blood from a broken syringe.

The light flickered. 'What the duce has happened here,' Kirsten gasped. 'His fingers and shirt are smeared with blood and it reeks of sweat.' She looked at the ink-stained chart. 'Stephen looks uncomfortable with his neck twisted in such an awkward position. He's far too heavy for me to move, I better call Bjorn.'

'There's no need, I followed you along the hall. You staggered as if you were sleep walking. My God! The doc's out cold. Paul's in the kitchen, we've been yakking on over Ian's blunder. Go and tell Paul I need help to move him. I can't lift him on my own. Stephen might lash out and clobber me in the face.'

'Here's Paul now. I'll make a jug of coffee.'

'How could anyone miss his thumping big clodhoppers on our wonky floorboards? Should've got the damn things fixed yonks ago, Kirsty. Things have been so hectic lately. I'll have a mug of hot coffee.'

'Coffee! Coffee! Who said coffee?' Stephen Jarvis stretched and yawned. 'I have mine strong and black. No sugar, thanks.' He looked at the ink-stained chart and scattered pages lying amid broken glass on the floor. 'Thank goodness my lecture notes aren't blood-splattered, or I would be in strife with the Varsity Dean.'

Picking up each sheet Bjorn looked at him. 'Poor sod's beggared, really stuffed.'

His wife lowered her head to hide a smile. 'Must you talk in slang. It sounds uncouth, Bjorn. The kettle's boiling. Is your crimped neck stiff or sore, Stephen? Mine would be after sitting in such an awkward position for hours.'

'No, I nodded off not ten minutes ago,' he yawned again. 'Once I transfer these details to a fresh page, it'll take me an hour to collate those fifty pages in your hand, Bjorn. I can't turn in, until I've taken Gigi's blood pressure and recorded her vitals on this chart. Why all the fuss?' Stephen flapped his hand. 'My stamina's like a rubber-ball, it'll bounce back after I have a cup of hot coffee.'

'Now you're talking rot. You need cocoa or warm milk. Not coffee, Stephen.' Behind his hand, Bjorn whispered to his wife, 'Put a drop of honey in it, and give it a good stir. He can't drink grog while on duty. I'll sleep in here tonight to give you a spell.'

She shook her head, and in defiance pouted. 'No, you won't. I've been sleeping on this divan. If Gigi awakens and needs to relieve herself, I'll be here to assist her. Your left knee's leaning on the old potty-chair.'

Jarvis's horrified look shocked Bjorn. 'Hey sport, are we on the same wavelength? Gigi never drinks to excess. Did some fool feed her a Micky-fin, or doctored her coffee with some toxic substance in the Yass café?'

'That's a million dollar question? The Canberra lab should ring soon with the results of her blood and sputum tests. If not, my receptionist will ring here tomorrow morning.'

Before retiring, Stephen Jarvis scrolled through his lecture notes on hallucinogens which included their effects on the brain and central nervous system. He read the lecture to verify its authenticity "All illicit drugs, whether artificially manufactured in illegal laboratories or natural have the power to impair and destroy a person's brain. Hallucinates, if taken indiscreetly over long periods cause memory loss and end in paralysis. The list of barbiturates and hypnotic substances around the world include tranquilizers. If taken in conjunction with hallucinates and alcohol they have the potential to cause paralysis and death. If in doubt or worried confide in your physician".

Tossing the lecture folder on a chair, Stephen checked his patient's vitals in the next room.

'Sorry, if I disturbed you Kirsty. Her temperature has dropped dramatically. We'll know by morning if she needs hospitalising. Her skin is clammy, but her breathing is regular. I think she's on the way to a full recovery.'

'What a relief, Stephen. I think she needs to relieve herself…'

'Use that pan. As I lift her, slip it under her bottom. The urine will flow naturally. I can't express my relief enough over her stamina and will to live.'

This done and the pan emptied, Kirsten recalled Kay Brown's promise to ring if Gigi needed her to collect Kendall from Goulburn in the morning. Keith and Kendall were schoolmates and shared the same room in the college.

15

5 am. The next morning: Ian barged into Jalna's kitchen and demanded to see his wife. Bjorn's raised hand warned him not to move a step further in the main hall.

'Everyone's asleep. That goes for your wife and mine. Back off, if you don't want me to call the cops. And keep well away from the house in future, or I'll whip the daylight out of your arse. This is your last warning. Your wife's staying here until she's well enough to walk unaided. Doc Jarvis is with her now. Now get lost, Ian. If I catch you hounding my staff again, like you did to Helen the other night I'll tell her son. Paul will run you in smartly. I intended to congratulate you for being the new part-owner of Abergeldie. Instead, I've decided that you are no longer an employee here, on Jalna.'

Ropeable over his decision, Ian's fist threatened Bjorn. 'What right have you to keep my wife here? I intend to speak to Gigi, whether you like it or not. Step away Svensson, or you'll feel the full force of my knee in your groin. If I leave, so does she. You've lauded the riot act to me once too often. There'll be a bullet in your RMB, before I leave this property. Be warned, it's not tagged with your name.'

Bjorn recoiled at the stench emitting from his gaping mouth, as Ian tried to forge past him. An outstretched foot prevented him from moving. The whip sprang into action.

'I never use this cat-o-nine tails on my horses or dogs. If you defy me, I'll wrap it around your grungy neck, until you plead for mercy. Lanky can drag your flayed body behind his horse to my roadside drum and leave you there. Later this morning Kirsten will transfer your accrued holiday pay and a fortnight's wages to your wife's private bank account. Now get off and keep off this property, Ross.'

6 pm. Bjorn knew Abergeldie, the sheep stud adjoining Jalna, was a sound investment for the Ross family. In the lounge room he looked, paused and looked again. What he expected to grasp was missing. Instead of making it obvious, he poured a whiskey from his own pride of prime Scotch and passed it to his foreman, Arthur Hancock. Holding up the cut-glass carafe, he again indicated to the women what their preference might be.

Kirsty took the initiative to confirm, 'Harriet, your doctors have advised you not to drink while on medication.' She frowned at her husband. 'Bjorn ask Bernie and Paul if they'd like a whiskey on this, the bleakest night on record for Yass.'

Instead he acknowledged Jarvis who stepped in the lounge room.

'Problem solved. Gigi's hallucinations were caused by a drug commonly known as LSD. The proprietor of the Continental booted two hippies out of his café for smoking marijuana and popping pills, just before Ian and Gigi entered and sat at their table. Crushed residue of a tablet ground into the floor has also been identified as LSD. I think she'd mistaken the tablet for her prescribed one. Where's Ian Ross now? He will confirm my suspicions.'

'Stephen, I saw him hanging around the coolroom breezeway as I walked in.'

'I ordered him to leave this morning. Why the hell's he back here? I'll soon find out.' Absolutely livid Bjorn roared. 'Give me five minutes alone with him. Where's that damn whip?'

'We now know how Gigi was drugged and by what drug. I'm not accusing Ian, or anyone here. It's imperative for the police to receive this information. The culprit may sell or give LSD or another insidious drug to teenagers or young kids in this district…'

D I Raddick interjected, 'Stephen, let me handle this in a diplomatic manner. I'm sure you won't mind either, Bjorn. I confronted Ian Ross the other night and he knows I won't tolerate his brutality or nonsense. I think he will confide in me, even if he's not the culprit. I also saw him near the coolrooms later that night, after I bought mum home from seeing a film in Yass.' In agreement his mother, Helen nodded from the kitchen as Ian Ross entered the house.

'When I challenged you yesterday Ian, you said and I quote: "I left the café because my wife constantly complained of a headache. Fed up with her whining, I then went to the pub to pay my dues". Is that right?'

'Yeah, a couple of hippies challenged me to a fight. This is a token of their brutality. My bruised knuckles. Gigi asked the girl for a glass of water to take a pill for her headache. Don't accuse me of giving her an illicit drug. It's typical of your breed. You cops are notorious for setting up innocent people just to gain notoriety to get a conviction.'

'You've seen my credentials.' Raddick again displayed his Inspector's card, badge and gun. 'Mr Ross, once you've signed this written deposition in front of Doctor Jarvis and myself regarding what you just said about your wife's tablet taking, you're free to go. Clear off this property and don't return. Heed my advice. If I hear one word of you showing your ugly mooch around here again, the local boys will throw you in jail. Leave now, before I change *my* mind.'

Ian didn't need a second warning. He shot through to his car, without glancing over his shoulder at the weapon dangling in D I Raddick's restless fingers.

Nobody felt like eating. Baked dinners remained in the warming oven. Bernie Staples refused a whiskey. 'I gotta be ringside before tonight's bout starts, so I won't stay thanks.' Jalna's foreman checked his watch. 'Abergeldie's a good buy. I wish they'd accepted my bid, instead of Ian's. See you at six tomorrow, boss.' A roving peck on Kirsten's cheek, a goodnight wave to Raddick and Jarvis, and he scurried out to his ageing Vauxhall.

Helen ate her meal in the kitchen. With her daughter nowhere in sight, she began to worry. Then she realised Jasmine and her niece, Rose were going to a dance in Yass.

Where is Ian? He pushed past me rudely. He's annoyed over some triviality. I'll soon know the reason why the boss checked their bunkhouse. Ross was in a vile mood this morning before they left for the solicitors. I think I know why! Gigi's bruised arms.

'Helen, I need to speak to you in my office.' Bjorn waited until Mrs Raddick wiped her hands on a tea towel. As he cut behind her outside the kitchen, his hushed voice held an air of urgency. 'I won't keep you long, but this is important.'

Sharply turning around, Bjorn almost collided with a belated guest, as he walked in the back door. 'Ted, go through to the lounge room. Kirsten's talking to everyone in there.'

He gestured to his housekeeper to remain silent. 'What I need you to confirm Helen, can be done in the kitchen. It's closer than the office.'

She placed two hot coffees on the table, and moved her daughter's Christmas cake, for Bjorn to sit down. 'Enjoy the dance tonight, Jasmine. Your cake should win hand's down. Don't forget the cut-glass stand. It's wrapped in a tea towel in that basket by the door.'

'Gee, the fruitcake looks delicious, Helen. You, and your sister Harriet and both your daughters all passed your exams last week with distinction.' Helen nodded. 'I lack the patience to do decorating or fine work of any description.' Bjorn knew the four women had spent a week creating waratahs, tiny wattle flowerets and flannel flowers of fondant and had dried each petal on beer-bottle tops supported on nails driven in timber.

'Kirsty's front garden is full of roses and native ferns. The reddish-brown boronia's with their green centres are her favourites. Now tell me Helen, during the day did you see Ian Ross prowling in or around the house?'

'Yes, I did see him take a bottle of whiskey from the bar. I didn't want to tattle, cos I thought he'd replace it. *You* don't *think* Jasmine or Rose would steal it? As you know Mr S, they only drink a wee drop of sweet sherry at birthday parties or here at Christmas.'

'Don't feel guilty for telling me. I suspected him all along. His wife needn't know. I'll tackle Ian when he's sober. Mr Hardacre, is his meal in the warming oven?' She nodded. 'Good, I'll tell him it's there. He's in the lounge room talking to my wife.'

Mrs Raddick understood. 'I didn't mean to dob Ian in or tattle-tattle, Mr Svensson.'

'You've done nothing wrong, so relax. I'm not blaming you,' Bjorn confirmed in an assured tone. 'We both know how deceitful Ian is. Don't worry, Helen.' He gestured for her to be seated then shut the door. 'I needed to know the truth.' *Tomorrow I'll tackle this dilemma fair and square in the middle.*

'Helen, have you moved my new bottle of whiskey. It's also missing from the lounge room bar? Why I'm asking, I only put it there yesterday.'

'You mean the brand Mr Ross buys?' It puzzled her, until she admitted seeing it.

'Thank you. Remember, you're not dobbing him in. Did he enter the house late yesterday, and if so at what time?'

'I can't honestly say, Mr Svensson. I saw him by the back-screen door after you ticked him off. I'd been down to feed the dogs and I noticed his grumpy look. The back verandah light globe's fused. Did you know...'

Damn, I meant to have the bulb replaced. I'll do it myself in a tick. It keeps slipping my mind. Bjorn frowned. 'Sorry, I interrupted you Helen. What were you going to say?'

'Last night I caught a glimpse of him again. None of your men have a massive build. Only Lanky, and I'd spoken to him near the first bunkhouse.' A finger pressed against her temple reminded her of the incident. 'Ian must've turned on the kitchen light, cos I'd switched it off moments before. When I poked my head in the verandah I saw a flashlight flick on and off. I could see him clearly. Not his face though. And I'm sure it was him.'

Bjorn leaned over a chair while listening to Helen. He chose not to say anything until she confirmed what he imagined to be the truth of his pilfered whiskey. *None of my trustworthy staff would intrude on our privacy without first making their presence known.* An idea flickered as Bjorn increased his span of intuitiveness. *I know Ian sneaked indoors the day Helen caught him perving on my wife. It occurred after she heard me abusing him for being a despicable swine for abusing his wife.*

A weird flash of memory crossed Helen's weary mind. *If I say too much, I fear retribution from him. Ian Ross seems to sense if I've spoken to his wife about his arrogance and rudeness. Gigi looked a little peaky this morning when I put clean towels and a fresh face washer on her washbasin. I wondered why the bloodspots were on her bathroom floor? Probably her special time of the month.*

Bjorn intruded on her thoughts. 'Nobody will know you've spoken to me. Keep your eyes peeled in future, and thank you Helen. I am indebted to you.'

Mrs Raddick stood to thank her boss. 'Mr Svensson, I trust you not to disclose what I told you tonight. I'd hate anyone to think that I'm a gossiper. A betrayer of secrets.'

10 pm. Helen finished tidying the kitchen and paused in the hall when she heard a muffled cough. 'It could be Ian on the prowl again. Is anyone there?' No answer, so she hurried to her room. Jasmine had put a fresh, cool nightdress on her mother's pillow. A delicately perfumed bar soap sat atop a clean hand towel and washer on the bed. In the shower recess two thick, absorbent white bath towels hung over its brass rail.

Slinking back to the bunkhouse to retrieve his father's gold watch and tan satchel, Ian, who lacked paternal instinct, scowled. *Why do I need another brat disrupting my lifestyle?* The aspect of a distinguished career now lay destroyed in a bygone era that no longer existed in this disorganised, humdrum and drunken bum's life.

Disorientated, he fictionalised a jealous fool weaving a spell in his dreams. The face of a beautiful woman he once spurned appeared when inebriated. Unnerved by her deep saffron eyes haunting him, Ian imagined he'd heard her soft whimper in the silent room. Madeleine had warned him twice tonight to keep clear of the house. In her misery she'd found peace in 1944, after being left destitute on the night of his staged suicide. The same night of the premeditated murder of a friend, his identical clone.

Thousands of times he'd tried to escape her constant taunting in his nightmares and took solace in the demon he'd grown to love. An inebriate's only friend…alcohol.

Now in the silent hallway he paused to think. *I will repel the bondage of love she holds over my subconscious mind. My strength to destroy evil will outwit hers. No, Madeleine isn't taunting me. In all probability it is her ghost. She professed to love me, and taught me the art of seduction. The army influenced all officers to eliminate fools. Wisdom coupled with knowledge is the key of combating an enemy.*

Ian remained in the shadows until Helen had retired. Silently he entered his wife's room. Kirsten was asleep in the chair. He suspected Gigi was pregnant. Reeking of grog, he sat on the bed. The force behind a flailing fist collected her chest. The pain was so intense that she clutched her left breast as he uttered about the death of an embryonic fetus. A sudden thrust from his other fist targeted her stomach. She groaned in agony.

'That'll teach you to disobey me. If you open your damn mouth, you'll cop another thump. Have you told this bitch you're pregnant?'

'No Ian. Nobody knows. Not even Doctor Jarvis has the faintest idea.'

'You won't get the chance to tell him.' Ian lashed out with his foot and collected her lower abdomen. Followed by a harder kick to her kidneys, which sent her flying off the bed. She landed on top of the drip-stand.

Lying in a pool of her own blood she pointed. 'Ian, I'm hemorrhaging. Wake Kirsten or call Stephen Jarvis. Don't leave me here. Call an ambulance, I should be in hospital.'

'They'll know soon enough.' As he slammed the door, Kirsten awoke.

'Gigi, what happened? Your legs and nightie are soaked in blood. Don't move. I'll call Stephen. Are you pregnant? If so, how far?'

She flapped her hands. 'How can I move, Kirsty? Yes, I'm three months. Don't tell Ian that I might lose the baby. He'll be furious.' She was talking to an empty room.

'An ambulance is pulling into the drive now. I caught them on their return from Yass, after being called out to a false alarm. They'll rig a new drip-line to the cannula in your wrists, Gigi.' Kneeling, Jarvis asked, 'Did I hear Kirsten say you could be pregnant?'

'Doctor Jarvis. I missed two periods, so I must be close to three months.'

'Have you seen a physician, or booked into a hospital?'

'Ian thinks it's a bit early.' She gripped her stomach. 'The pain is getting worse.'

'You'll be in hospital soon. Gigi, these ambulance officers will take care of you. I'll see you there. Keep breathing deeply for it will help to elevate the pain.'

'My toilet bag and night clothes…'

'No need to worry, Kirsten's taken care of everything. Your packed case is on the ambulance trolley.'

'Darling, I'll pop in to see you, once Stephen lets us know what happening and how you are feeling. Keep thinking of Kendall. He'll be home soon. I'll bring him to the hospital, he'll be anxious to know how you're coping with the pain.'

10am: Gigi had just signed the consent form in bed, a precursor to going to theatre, when Kirsty and Kendall arrived. Although his eyes were swollen with tears, he didn't mention his father.

'Mum, are they going to operate?' He had an idea she might be pregnant, but didn't know if she would lose the baby. This subject he'd studied in a private seminar at college. Now his astute mind formed a different picture. One he dreaded.

A glimmer of hope flashed in his downcast eyes, and as Gigi smiled he clasped her hand. 'Mum, it's good to see you smile when you're in so much pain.'

'This form is only a precaution. Don't worry, son. The specialist and his team are professionals at solving mysteries. I'll be in the best of hands.'

'Kirsty filled me in how it happened. I promised her and BJ I wouldn't blubber or upset you. You will be okay, if you don't insist on coming home too soon. Mum, I'll come in again this afternoon.'

'No son. I need to rest. Don't worry your father. He's busy selecting the furniture for our new home. And don't go worrying the people on Abergeldie. They're due to leave there today. Here's the trolley. Does this yucky-blue cap suit my complexion, Kendall?'

'Mum, what drug have they given you? You're high as a kite in the wind.'

'Kiss your mother and give her a hug. The nurse and a wardsman are ready to wheel her to theatre.' Kirsty stepped back to let Kendall stand closer to the gurney.

'Love you heaps, Mum. Kirsty is going to run me home now. I'll miss you and that funny mesh cap. It's looks cute on your wobbly noggin.' Sniffing he walked away.

'I'll ring the hospital at two. We'll be anxious, until you return to the recovery ward. Bye darling. Kendall will be fine with us. Bjorn has a special project lined up keep your son's mind and fingers occupied. BJ sends you his love, Gigi.'

Time marched around the clock five times before a splattering of news came over the phone.

'The result of her blood test is what? Stephen that's incredible. Why didn't she complain of pain long before she collapsed? Gigi always confides in me. She will recover? No, it's not what we all expected. Yes, I will be tonight. Bookwork never ceases on our homestead. What time? Dinner at six. Kendall's working in the stables with Bjorn. They're fixing the cane-latticing on a chair. No. It's a surprise for her, not me.' Hanging the phone on its wall cradle Kirsty sighed.

5am. The following day. Saddling her horse in the stable, Kirsty looked to see a shadow hovering over her head. 'What are you doing here? My husband and Paul Raddick have warned you to keep clear of Jalna. If you don't leave now, I'll call Lanky. He's feeding Muffin in his stall.'

'Useless phoning. My wife, where is she? I rang the hospital and got no reply. The receptionist refused to put me through to her ward. Some embargo on me speaking to Gigi. If you don't or won't tell me, you leave me no option. Must I force my way into the house, to tackle your bastard of a husband?'

The smirk on Kirsty's face grew bigger.

'I doubt if that will be necessary. If you move an inch, my bodyguard's hand will wrap his whip around your neck.' The sharp sound of cracking leather-thongs was deafening in a confined space. The butt of the whip handle struck Ian's kidneys. His knees buckled. He groaned in agony.

Two hands with brute force gripped the collar of his plaid-shirt and wrenched his neck sideways. Bug-eyed, he stared up into deep-brown eyes flecked with gold.

'Where do ya want me ta ditch this bag of rubbish. He ain't gonna hurt his missus, not while I'm working on Jalna.'

Kirsten's thumb angled towards a bale of hay. 'Sit him on that, Jamie. He won't budge if you press the whip handle hard down on his shoulder.' She parked her bottom on a milking stool. 'Right Ian. Your wife will be in hospital for two weeks. Her illness was not what we all imagined. I refuse to discuss it now. Kendall will be staying here on Jalna until he returns to school next January. He will never set one foot on Abergeldie's soil while you're there. I'll bring Gigi over, after her recuperation period in three weeks. If Bjorn or I catch you within cooee of our front gates before then, we won't hesitate to ring D I Raddick. I know Paul challenged you, after you threatened his mother. And Harriet objected to you threatening her. Jamie, make sure he leaves this property.'

A nod accompanied his broad smile and Jamie's dark eyes gleamed while listening to her abusing Ian Ross who accepted his dismissal with a smart growl. Proudly tucking his whip under one arm, the native backtracker gave Kirsty an Aussie salute. He walked his horse up to Jalna's gates with Ian beside him, a rope tied to his wrist. The boy then waved him off and with a warning he then rode across the lower paddocks to see his elderly mother. Peggy Ludawalla, ready to bring in the washing, eagerly greeted her son. Dismounting, Jamie lowered the forked clothesline prop and moved the washing basket within his mother's reach.

'Jamie, put the whip in your bedroom with your other trophies. Ya know the boss hates whips. Do it now, boy. Then carry these things inside. It smells like rain and there's a willy-willy sweeping them hills free of tumbleweed. Hurry there now, or the missus will growl. Her best sheets and towels are in that wicker basket.'

16

On her return from the hospital, Kirsty's car pulled in through Jalna's gates and Kendall pointed to the RMB.

'Pass me the key to your mail drum and I'll see if the postman's been.' She handed him the key. 'Leave the motor running Kirsty, I can reach the drum from here.' He pulled out a pile of mail and passed her two official looking envelopes. 'There's a letter in this lot for *him*. Why would anyone from Liechtenstein be sending *him* a letter?'

She flicked through the pile.

'Kendall, here's another. The franking stamp is from London. This small letter is from Dresden, in Germany. Replace that rubber band and drop the key and that bundle of letters in my purse.'

As they entered the house his mind lingered on his mother's pre-theatre comment. 'How come mum joked about the yukky blue cap when she was in so much pain?'

'Perseverance to survive. All her adult life Gigi's had a spirited fortitude to survive under the harshest of conditions. She thrives on fighting a challenge and usually fends off her opponents with witticisms. She can be clownish, or buffoonish, yet still be serious in the most devastating or deathly confrontation in an argument.'

'Thanks for filling in the blanks. The peculiar thing is I thought she was hallucinating, because of some drug they had given her, before she went to theatre.'

'Kendall, do you remember the first time I visited you in Paris and you offered your cold cocoa to me? It tasted like heaven after my long trip from your home in Evesham.'

'Yes clearly. Why? What an odd thing to ask. I was around three or four, I remember saying, "They is here, Mummy". You laughed because I'd said *is* instead of are and gave away their secret. We were supposed to surprise you, not the reverse. Mercedes took me to her room. I still have the nursery rhyme book that she printed my name in.'

'Well, I think you suspect the truth of your Mother's so-called accident.'

'I sure do. That swine kicked her in the stomach. It reeks of his nasty nature. I hate him. If he lashes out and belts mum again, I'll threaten to whip his backside with his belt. It won't lower your standards Kirsty, by saying what I call him. He's no longer a relation of mine. On my eighteenth birthday, I'm going to have my surname changed by deed pole, *Auntz.*' They both smiled. 'Two hot coffees coming up. Milk in yours, right Kirsty.'

An intuitive lad for his age, Kendall sensed his sibling was either aborted or had died in his mother's womb. Within the hour they would know the truth. Restless, he changed the subject to a more menial topic of interest.

'Before mum and I delved into her colourful cap, I wanted to tell her the results of my school exams. It can wait until she's out of danger and leaves this hospital. A's in mathematics, trigonometry and biology. I excelled in English lit and history. Although I can't boast of scoring a B in geology. I flunked sport, especially tennis. I dislike all grades of footy. It's boisterous and reminds me of neolithical creatures who battled to survive when dinosaurs ruled the earth and uninhabited lands. My favourite subject is astronomy. I study the planets of our solar system keenly. Kirsty, the Aurora Australis and the Aurora Borealis both intrigue me. The fluting and intermingling of their green, blue and purple curtains of beauty look like sea anemones dancing in a calm ocean, or amid turbulence in stormy weather.'

'My! Quite the young professor, aren't you? A close friend once called your mother a philosophic professor because of her intuitive nature. Now it appears there are two in one family, Kendall. It will give her a thrill to hear how well you have progressed at school this term. No, all through the year. God alone knows, she needs some contentment in her droll life. Life hasn't been all sweetness, or beer and skittles for her. Your father is a different kettle of fish. He does not rate a mention in my books.'

'Nor in mine, Kirsty. He is a beastly hypocritical bastard.'

2 pm. Doctor Jarvis removed the blue theatre gown and dropped it along with his mask in a container of disposables. 'Your mother's in recovery. You can go in now, Kendall. Stay only a minute please. Gigi is incoherent and she may not recognise you. The Intensive Care ward is further along this corridor, lad.'

Kirsty was a little hesitant to speak until Doctor Jarvis raised his hand. 'I know what you intended to ask me. Gigi wasn't pregnant. The specialist battled to save her life. An ovarian cyst and a smaller one in her uterus were removed. Because of her severe hemorrhaging and infection his team were forced to do a hysterectomy. This is a merciful blessing. The ovarian cyst was bigger than your fist and riddled with tiny clumps of knotted veins. In three to six months she would have suffered immense pain and may've died.'

'Stephen, I really thought she was pregnant. Gigi has the constitution of an ox. Like you, I know she will accept the news of her hysterectomy with the grace of a veteran.'

'I have the same faith in her recovery. It will be three months before she should lift anything, or raises her arms to a great extent. Sweeping and carrying heavy items are taboo. The underlying sutures in her abdomen are dissolvable. Unlike the surface sutures of fine gut. One of the nurses will remove them in eight to ten days, Kirsty.'

'How long will she be in hospital? I heard Kendall ask a nurse if he could visit his mother again tomorrow. On her discharge, she will be staying with us on Jalna. Your hotel room is permanently booked, Stephen.'

'Tar muchly. One of Bjorn's wacky sayings. A week, may be longer. It depends on her recovery after these transfusions. Her blood is AB positive. A rare grouping. If you don't mind, I must finish my rounds in ward three. Then I'll collect some fresh clothes from home. God willing, I should be at Jalna by six. Look after the lad, Kirsty.'

Approaching the boy Jarvis shook his hand. 'Kendall, you're a credit to your mother and she treasures you. I can't foresee a reason why your visit will be objected to by staff. Gigi will need congenial company to overcome the trauma she suffered prior to entering hospital. I'll see you both tonight'. Jarvis touched the boy's arm in a comforting gesture, then with humped shoulders and a furrowed brow he plodded down the corridor.

Kendall hurried on ahead to do the gentlemanly thing. 'Hop in Kirsty, here's your keys. You dropped them while talking to Doctor Jarvis. What did he say about Mum? Hope she can be discharged from this ice-cold morgue and soon.' Kirsty frowned 'You knew what I meant, Auntz. Be honest. Doesn't this hospital remind you of a bleak mausoleum?'

'Not really. But then I've never wanted to visit a pharaoh's crypt or tomb in Egypt.'

'Did Jarvis tell you *when* my mother will be *discharged*? I must know, so I can make sure *he* won't be on the scene. Need I spell his name in full? Initials will do. IMR. And I don't mean your roadside drum. If I ever catch the bum stealing her mail, I'll shove him in it and then throw the key in Abergeldie's cesspool.'

'I'm curious as to how you know they have a cesspool. We do. But do they? You do realise what produces the gas that dissolves sewage?'

'Yeah. We studied it at school last year. A pig or a ram's head full of mag ...'

'Don't *go* into the *details*, Kendall. I don't want to be sick on an empty stomach.'

Kendall imagined what complications may arise in the immediate future, and gave it extensive thought before admitting, 'In a trillion years I *will* own my own horse, a horse stud and ...'

'Why wait a trillion? You have the potential to own the world, if God grants you the wisdom. It seems you do not know what day it is, by saying such a foolish thing. The book on your lap? It looks interesting.'

'A study on heart conditions and colic in unborn mammals. I think my future lies in being a vet. Or I may further my studies in science. Scientists make a fortune in studying the universe and distant galaxies. I find the heavens fascinating. The mists remind me of the eucalyptus haze which forms above the Blue Mountains, on warm summer days. Variances of blues and purples like the Aura Australis. Magical designs of the cosmos in all its beauty, creations of nature have inspired science devotees for eons. Evolution is a wonderment to inventors, who strive to make the earth worthy of us humanoids. We live in an ugly world in which evil devours or destroys everything, or ruins the things we value most. But they are only monetary things. Life and living it well is more important.'

'You sounded like a dedicated young professor, Kendall. And I could listen to you all day romancing about unblemished distant heavens. Only we need a bite to eat and a cup of strong coffee in a café on the way home. It seems the contagion of time is the evolver of iniquity. I believe you want to buy a horse. An Arabian stallion or a Prussian mare, that you may give a historical Russian name. My psychic instinct tells me perhaps a dapple-grey, or may be a magnificent, sleek black filly.'

Her vision inspired Kendall to think: *My favourite Auntz you may not know it and it will be my secret, but you have awoken the phantom of my dreams. I shan't tell you now. Not until I visit a special presentation at an equine sale with your husband.*

'Kendall, pass me my handbag please. It fell under your seat a moment ago. The car is running on fumes. This bowser is the closest to its cap. Here's the key, be a sport and unlock it for me. Then I need to pay for the petrol. I won't be long. Your mother could be coming home late this afternoon. I can't be sure, not until I've spoken to Steven at noon.'

'Yippee! What bonza news. It'll be beaut to be with her, without him being there.'

2pm. It turned out to be a low-key birthday for a young man whose heart was set on seeing his mother walk unaided in Jalna's front door. Feeling isolated in a roomful of people, he escaped a barrage of questions and fled to the main hall, where he collapsed in chair.

'Mum, oh Mum why have you forsaken me? All my prayers have gone unanswered.' A stream of cool, salty tears ran down his hot cheeks as both hands covered his eyes, closed in anguish.

Suddenly a friendly hand gripped his left shoulder. 'Come on lad. I have something to show you in my room. Rose, Jasmine and I have tried to console you in your misery. We're feeling miserable like you. The news touched my heart and Harriet is inconsolable. There is a brighter side to this, Kendall. Take a deep breath and look behind you.'

'Please Helen, let me collect my thoughts. I loved my mother and now she's gone.'

'Gone where?'

Startled, he looked around. 'Mum, I can't believe you're here? I phoned the hospital and a lady told me to wait…'

'Kendall, I was there and I heard her. Kirsty had gone to bring the car around to save me walking far. What you heard, were two men discussing why their mother was in the morgue. Son, you assumed they were referring to me. Give me a hug, and no more tears. I'm home and happy birthday, darling. Once I struggle out of these damp clothes, we can celebrate by enjoying Bjorn's barbequed meat, onions and salad with ice-cold lemonades.'

17

.30 am next morning. Clad in night attire and with a glass of water in her hand, Gigi sauntered to the bunkhouse to see what Ian had left behind, after absconding the previous week. The hot water wasn't on. He'd switched it off. Retying the sash of her robe she leant on the kitchen bench to rest. She noticed an empty beer can and a whiskey bottle with liquid dripping on the timber floorboards. 'He must've drunk himself silly. Stupid man! Well, I won't be his slave now. I'll clean up that mess after breakfast.'

'Oh, no you won't, Gigi', Helen said tossing a bucket of disinfected water over the floor. 'One of our young lassies can clean the benches. Has Bjorn, or Kirsten told you that your husband has accepted our offer to work on Abergeldie? Rose and I are moving in tomorrow, along with three local girls to do the housework. Jamie had also offered to spend his free time working in the stables. They've altered it to house your workmen, rouseabouts and the casual shearers. Bjorn's men are drenching and crutching your sheep this morning.'

'Will wonders never cease? What made Ian change his mind? At one stage he ranted and raved that not one of Jalna's men, including Bjorn would be welcome on Abergeldie.'

'My son did. Paul took a dim view of the way Ian spoke to me. That's all I can say for now. I think he may have given Paul's warnings a lot of thought. Before he left, Ian hired two elderly gardeners to create a magical picture of beauty for you. The designs you threw in your bedroom bin, I gave them at his request. One bloke asked Bjorn for a wee suckling of our cedar tree, and he planted in your front lawn this morning.'

'Miracles! I'm itching to walk through the homestead. I've only seen photos. The day we bought Abergeldie, Ian promised to take me to inspect the house and property. Instead I landed in hospital. Helen, I feel on top of the world now. Fighting fit. I hope your sister Harriet will be our cook and my mentor on Abergeldie, as she promised.'

'She and Rose have hung your bedroom curtains and bought a matching bedspread. The Edwardian floral garden pattern you admired in a magazine last week. There's a fluted-bell lightshade above your bed. And the bathroom has delicate green-tiled walls and a new washbasin. I shan't tell you what towels. That's their surprise.'

Trying to keep her balance in the cool, lilac scented air Gigi ambled up the path to the pool under the corrugated-iron tank stand. She bent to sip the cool water in cupped hands. It diluted the bitter taste in her mouth. A rancid smell of stale beer lingered in her nostrils.

'Once the sun rises and burns off those dark of clouds, it'll be a hot, steamy day. Sit here in the breezeway Gigi and I'll bring you a breakfast tray. There's a pile of magazines on this chair. Did you notice the new wicker-work on your bedroom chair? Before things went haywire, Kendall intended to give it to you. He and the boss rethreaded each strand until your old chair was restored.'

'The tuckpointed swans and their signets on a blue lake with a sky of pink and white reminds me of a summer's day in France. A splendidly woven tapestry and quite valuable. The contoured seat surrounded by delicate green and its padded back have been beautifully restored. How and by whom?'

'I shan't tattle of another secret. It will spoil his surprise. Kendall's eager to tell you. Now you must wait, Gigi.' The reflection in Harriet's eyes, imaged the glorious sunrise as it peeped through a break in clouds and looked like puffy pinkish-white fairy-floss.

Skimming through one of the magazines, her fingernail caught on a small envelope tucked in the front page. Gigi recognised the pay packet. 'I'll give this to Kirsten.' She could feel something firm in the envelope. *No, it hasn't the crispness of money.*

6am: Awoken by a weird sound Kirsten approached her kitchen and saw a silhouetted shape against a backdrop of the breaking dawn. Hurrying to the back wire door she called, 'Is that you Bjorn? Why are you standing in that cool morning air?'

A faint whimper disturbed the peace. 'Kirsten, it's me. I feel a little dizzy.'

'Gigi, you're shivering. Come through to my office. The heater's on and we won't be disturbed.' She grabbed a woollen shawl from the verandah stool, and wrapped it around her bleak shoulders. 'Darling, you could have caught pneumonia out there. Why didn't you buzz me?' Kirsten remembered the buzzers in the bunkhouses needed fixing.

'I found this envelope in a magazine. It might belong to one of the drifters, or Bjorn's shearers. It's not Ian's. He doesn't read women's brochures. If Kendall hears him abusing me, he might thrash his father.'

A warm cocoa warmed her through and settled a palpitating heart. 'Thanks Kirsty I needed this.' Sipping it slowly, she pointed to the flower on the desk, then noticed one on a wrapped gift. 'Those wooden roses are exquisitely beautiful, where did you get them? Their perfume is quite delicate.' Gigi felt a twinge of pain in her stomach to support it. The sutures were tightening.

'Look at the huge cedar in our front lawn. Bjorn planted a seedling of his uncle's tree, before we moved in here. It's a wonder, with your intuitiveness, that you haven't noticed these roses on the ground and how its fine-leafed boughs droop like a weeping willow.'

'Well! Not really. My mind's been on other things lately. Kirsten, I don't feel very well. Doctor Jarvis promised to remove these retched stitches today.'

'Stephen's on his way. If all goes well, he should be here soon.' She blinked twice. 'That's his car horn tooting now. A signal that he's pulling into our drive.'

'Thank heavens he's here. I'll pack my case then will you drive me home this morning? Ian's vile temper seems to have mellowed. Why he drinks to excess is stupid. I don't want to leave him, Kirsten. We own Abergeldie now, and I'm anxious to see my new bedroom and the bathroom. Ian says he's going to sleep in the back bedroom, so he can work until all hours in his den. Can I give him a buzz?'

Kirsten's frown softened. 'Darling, you don't ask to use our phone. Use this office one. Don't worry, I won't shut the door in our doctor's face. Come in, Stephen.'

Gigi settled down to hear the report of her tests from Jarvis. 'I'll remove those sutures in your room, Gigi. Before I do, I've devised a diplomatic way to speak with your husband. I'll warn him that you need a lot of care. You've suffered an enormous amount of stress. I may suggest to Arthur Hancock or Berny Staples to keep an eye on Ian. I know they'll both be working on the lower fences this week with your new manager, Tom Marden.'

Without answering, Gigi dropped the mysterious pay envelope in Kirsten's office drawer then shuffled to her room. Catching her before she entered it, Stephen offered to help her over to the bed.

'I can manage on my own. Give me a minute and then come in, Doctor Jarvis.'

Kirsten heard a snippet of their conversation and intervened. 'Forget this doctor rot. Gigi, he prefers his close friends and patients to call him Stephen. He's offered to pop in daily to Abergeldie to take your blood pressure and to access your health. For a busy man, I think you should be appreciative of his offer. Call him Stephen and you'll get on fine.'

'Okay, if you insist. He may object. Still, ask him to come in Kirsten.'

Jarvis described the removal of her sutures and revised his previous directives on her case-notes. 'Do you still have the copy of her previous injury report, Kirsten?'

'Stephen, can I speak to you privately?' He nodded and followed her to the bedroom door. 'That report no longer exists. I can't be sure if Ian stole it, the day he caught me naked in the shower. I have searched my office and the den. The copy is missing.'

'It's negligible.' He smiled walking back to the bed. 'I'll transcribe her notes to mine in the surgery. You should rest until lunchtime, Gigi. I'll be at the hospital if you or Kirsten need me. It is necessary for you to keep off those swollen feet. Your wound is saturating and I put a sterile dressing over the suture line. The medicated tape should hold in place until tomorrow. No showering though, dear.'

'I'll be fine in the morning. Can I please go home, if I promise to rest on my bed, or on the divan? Ian's expecting me Doctor, err Stephen.'

He smiled, collect his medical case. A minute later his car pulled out of Jalna's drive.

5.30am Ready to eat his breakfast Bjorn scowled on hearing a rumpus 'See who the idiot is banging on the back door, Helen. If it's his nibs, tell him I'm busy.'

'Mr Ross wants you. What message will I convey, Mr Svensson? He's not angry at all. The opposite, really.'

'Serve our ham and eggs up please Helen, while I speak to him. I know why he's hanging around here at this hour.'

'Where's my wife? I've come to take her home. The house is clean and tidy.'

Doctor Jarvis, who was leaving Gigi's room glared at Ian. 'I will drive your wife over to Abergeldie. I've just given her a drug to counteract this infection. We'll follow you there in ten or so minutes.'

Ian collected her luggage from near the back door and nodded to Helen, brewing a fresh pot tea. 'The boss will appreciate this, instead of coffee. Keeps the cogs ticking, and us all in a good mood. Would you care for a cup, Ian?'

'No, I'm in a hurry.' He acknowledged Bjorn with a side nod. 'I've left the front door ajar. Gigi, you must take care climbing our front steps. We don't want you to have another fall, or an accident. You're quite unsteady on your feet this morning.'

Ian's suave palavering didn't wash with either Stephen or Kirsten who looked at their departing guest, and covered her smile with a handkerchief.

Bjorn looked at Ian then a bucket and mop by the back door. *I should've made him clean up the rotten mess he left in their bunkhouse. My staff had to change, rinse and soak his stained sheets and bag them. I should've docked him a week's pay. God help his wife if he drinks to excess on Abergeldie. He reckons he's on the wagon. I doubt it. Ian's a liar and a drunken bum. How can a leopard change its spots? That bastard won't alter his gut-swilling habits.*

'Please leave now, Ian. I want to drink my coffee in peace. My crew will be in for their pays soon. I may come over later to see if Lanky and Tom Marden have mended those wonky fences. I'll bring a bottle of freshly brewed ale.' Bjorn then whispered to his wife. 'It's non-alcoholic. He won't know the difference, Kirsten.'

'Want to bet me five quid? He will. Ian's no fool. Although, I must admit this batch tastes more like real hop-beer. Quite mild in fact.'

Things progressed well for Kendall during the first week of his third term break. Bjorn and the boy spent the rest of the day at the Canberra yearling sales. It paid dividends for Jalna's boss, who bid on four yearlings, three bay geldings and a dapple-grey mare. They moved out of the direct sunlight to a shadier spot under a gum tree. BJ's prized buy was the last roan to strut around the ring.

'Boy, will I be glad to get off my feet. These boots are killing me, Kendall. Guess you're feeling a bit weary? Let's grab a bite to eat at the bar. I'll shout you a hot-dog, or a sandwich and a coke.'

With both hands bracing his hips the lad sniggered.

'No, make mine ham, lettuce and mustard topped with mayo. I love the mayonnaise that Harriet and her sister Helen made with fresh farm eggs. The stuff you buy in shops tastes yukky. Buy me an ice-cold, apple cider please and it'll make my day. Are we heading home soon, BJ?'

'Yeah lad. Lanky and Bern are about to put my two yearlings in the double-float. The other horses are ready to roll, in their floats. Good buying day, aye mate? Thought I might miss out on the first bay gelding. Someone upstairs must've blessed me.' Bjorn kicked a pebble. 'Phew, it's a cooker. Let's shelter in the shade of this willow. It should only be ten or so minutes now, until we're on the road and heading home to Yass.'

'Yep, I am a bit weary. It'll be good to soak in a hot bath followed by a cool shower.'

'I meant to ask you Kendall, which of my equines caught your attention?'

'The deep roan. His immaculately groomed pelt reflected a deep chestnut in the sunlight and his hind-hocks looked ebony or black. Why Bjorn?'

'Can you keep a secret? And I'll tell you.'

'Why the riddle? Yes, I can.'

'Guess whose horse he will be? But only if he's given a Russian name.'

'He's mine BJ? The roan cost you well over ten grand. Did you hear me say I'd name my horse Potchkin? Mum's going to call her stallion Wallaby Downs. That's if she can ever afford to buy one. It could be donkey's ages before our stud is profitable.'

On their arrival home Kendall looked disheartened as Kirsten greeted them near the stables. 'Is Mum still over on Abergeldie? She promised me this morning she'd be here.'

'Bjorn, let Kendall help Lanky walk your horses down from the first float. Then come inside. I found something that might interest you. Don't say a word to the boy. I'll be in my office.'

'Why? I need a hot cuppa, not to see something you've dug out of the files.'

'Don't argue, Bjorn. This is important.' She stomped off in a huff.

Thundering down to the office his voice took on a voluminous roar. 'Kirsten, I can't afford to waste time on some frivolous motion of yours. The men are waiting for me to unlock the damn saddle and gun rooms.'

Their door closed. 'Read those and then look at the photographs. I don't want to hear a peep out of you until you've finished. Do you recognise either of the women? I can't recall seeing them, in or around Yass. The tall, dark haired lady in that sleek black evening gown could be a mannequin or a model. Look at the tortoiseshell combs tucked in her French roll. A while ago Gigi mentioned she'd found a comb with the initials K M in Ian's cuff-link case. Now these snaps turn up in a pay packet. The number I wrote on the top corner is missing. Do you have any idea whose it might be?'

'No, Kirsty. It is a felony to tamper with the mail. All these letters are addressed to M. Kleinhardt. The franking on this one reads Liechtenstein. And this folded envelope was franked in Dresden. Lock this entire bundle in the bottom drawer of your desk. There they won't be tampered with or lost. How did you come by them?'

'The young painter redecorating our main hall found them. They'd fallen behind the antique umbrella stand. Immediately Ian sprung to mind. If we approach him, it'll upset Gigi. The last thing she needs is another confrontation. Drop everything in this drawer and it *will* remain locked. There's nothing important in that drawer. All our receipt books and ledgers are in the filing cabinets. The key I keep in my bedroom wardrobe. It's safe there, if I'm away.'

Bjorn had noted the franked dates and places of each letter. He was, however, mystified by the shredded photo of a young woman in a floral dirndl. Her soft, gold-braided hair resting on the nape of her neck intrigued him. 'Why would anyone destroy a photo of a loved one or a friend? This is mysterious, a real jigsaw. All jigsaws create a mystery and bring to life romantic, or intriguing pictures.'

The yellow-paged letters and photographs were forgotten. A more important issue loomed on a dark horizon. An unwelcome drama was emerging of hate and revenge.

4 am Saturday morning. The Svenssons were awoken by a loud ring of the hall phone. Bjorn yawned, jumped out of bed and answered it.

'What did you just say Harriet? Gigi is missing and her bed hasn't been slept in. You've searched the house and she's not there. Where is her wretched husband?'

'Mr Ross is asleep in his den. I think he's been drinking all night. There's an empty bottle of wine on the floor and his head's resting on the desk. His bathroom smells of vomit. I mopped the floor and then rang you. I don't know what's happened to Gigi.'

'I'll be over, as soon as I'm dressed, Harriet. Calm down and put the kettle on. If Ian's been drinking to excess, he'll need sobering before we can make him tell us where Gigi might be. Kirsten's gone to ask Jamie to saddle both our horses.'

4.30 am. On entering Abergeldie's homestead Bjorn found the owner plastered. It enraged him. He needed Ian sober for him to be fully cognisant of the message to be delivered. Sobriety of his neighbour was important and urgent. And it was important to make him comprehend how a man in his position should behave. No sober or decent man would treat his family this shabbily or so degradingly that his wife constantly lived in fear of her life. Neither Gigi nor Kendall appreciated a bully living under their roof. *Kirsten will be bitterly mistaken, if she thinks there might some good in this bludger. I'm damned if I've seen a decent trait in him.* A passive and compassionate man, Bjorn didn't believe in violence. Under extreme circumstances he had no option than to make the Scotsman see reason.

The brewed pot of strong coffee served its purpose. Before insisting that Ian should ingest more coffee. Bjorn dribbled a concoction called the "Hair-of-the-dog" down his neighbour's throat. Its potent properties would bring any drunk to his senses.

In a frenzy Ian tried to fight his assailant who was trying to force something down his gullet. A smell permeating from the kitchen reminded him of a similar drink he had enjoyed last evening. Even so, it took time and all Bjorn's patience, plus a great deal of strength to accomplish this feat. Ian spluttered, kicked and, on tasting the bitter brew, tried to push his protagonist away. Svensson persevered. Once Ian had swallowed the lot, Bjorn made fresh coffee, and trickled it down Ian's throat. When stable enough to stand Bjorn walked him around the house. After half an hour the farting wind-bag resembled a live human being.

While walking Bjorn tried to discuss things in a rational manner with him. Ian listened. But none of Bjorn's ideas made sense. It took ten minutes for the suggestions to penetrate his sozzled brain. The motive behind Bjorn's orders lacked commonsense.

'Now you're sober, tell me where you left your wife last night? I'm sure you had a hand in her disappearance. Her saddled horse found his own way home. If you don't tell me the truth, I'll phone the police. Then they'll throw you in the slammer.'

'She went riding up in the west top paddock around dusk. That's all I friggen well know. Go to hell, Svensson and get off my property and I mean now.'

Bjorn was ready to jump on his horse tethered to Abergeldie's verandah post when Ian began to heave. A rational and quick thinker, he grabbed the hose and passed it to him. 'Right, now flush away that disgusting mess. And watch where you're pointing that damn hose. I've had my shower this morning and I don't need another.'

Finished hosing down the verandah Ian then turned the hose on his neighbour.

Agility paid dividends. Having anticipated this move Bjorn ducked and copped a slight drenching. In retaliation he turned the hose on Ian. The stench of his saturated underclothes stung his eyes. It even affected his nostrils. Bjorn heaved.

'Now, get into that damn bathroom. And then don't leave your study. I'll buzz Harriet every ten minutes, so don't think of bashing her. There's dried blood on Wallaby Down's saddle. Your wife's blood. Now I'm calling the ambulance. Then Jamie, Tom Marden and I are riding up to the south paddock. Not the northern one. You're a bloody liar, Ross.'

With a firm hold of Ian's neck, he pushed him inside and threw him in the shower. The blast of hot water revived the drunk. Alert to a degree, Ian disrobed and washed his body free of his own vomit and mud. With only a towel draped around his waist he flopped on the bed. His soiled underclothes lay in a heap on the bathroom floor.

6.30. Bjorn handed his walkie-talkie to Harriet.

'Don't you wash his putrid clothes or clean that bathroom. I'll come back and make Ian do both. Press this button to buzz me. The green button is to speak and the red to answer. Now call the local ambos please Harriet, and tell them my men and I will be up in the east paddock searching for Gigi. Don't buzz the coppers yet, thanks Harriet. I'll let you know over the blower if we find her alive.'

2 pm: Abergeldie's top-off-side paddock. Suffering from hypothermia and with frostbitten fingers, Gigi was unconscious when Jamie found her icy and mud-caked body hidden under branches. The loud crack of his whip whistling through the still air coupled with his cooee alerted Lanky and Bjorn. With the ambulance following close behind their horses it reached his unconscious neighbour before a storm broke over the escarpment.

Later that week Doctor Jarvis collected her from the hospital and drove her home, much to the relief of Harriet and Abergeldie's entire staff.

18

Sunday 12 December 1965 - the morning prior to Kendall's 18th birthday.

Bjorn Svensson looked mystified as he walked from the dining room.

'Why the puzzled look on your face, boss? You don't often scowl. What has your pet aversion done now? I gather Ian Ross is on the breakfast menu.'

'Bernie, don't ask. I'm ropeable with that ignorant prick. He came in here rambling nonsense yesterday and now I find the magnum of French champagne is missing from the bar. It could only be him who took it from that lower shelf. None of my staff would think of pinching it. Kirsten will hit the roof when she finds out. It was Pierre's parting gift to Gigi, before he flew back to Paris in September. We kept the magnum, plus a bottle of cognac here, so Ian wouldn't see them. He wouldn't break the seal, knowing I'd discover the truth. Lately he's been swigging anything with an alcoholic content. How can we face Gigi? She promised Pierre to open the magnum today on Kendall's birthday. She'll be heartbroken.'

'Crikies! I'm stonkered to think of an answer, boss. I know he's a deceitful bastard, but to steal two bottles of grog Ian must be loony. You were going to uncork the magnum at the boy's party this arvo. Why the hell would Ross pinch a magnum? He couldn't guzzle the lot in one hit. Well, if he's inebriated the idiot might try.'

'I'll challenge him later, once our guests have left. I refuse to let Ian or a damn bottle of champagne spoil Kendall's special day. It is a rare vintage of champagne.'

'See you later this arvo, BJ. I better get going, or the missus will be snaky.'

'Be back here by six, or you'll both miss the barbeque, the delicious cakes and tucker, Bernie. It looks as though my lunch is ready. What's on the tray Kirsty?'

'Fresh ham and salad sandwiches with a smear of mustard. We were going to celebrate Kendall's passing his final exams with honours in five subjects. Bjorn, have you forgotten, we promised the boy he could uncork the magnum at Christmas?'

'We can't celebrate a damn thing now, Kirsten. How's his mother this morning? She looked pretty crook at six, when I took her a cup of tea. I'll kill Ian if he keeps belting or chastising her over insignificant things. You sound her out, do it tactfully. If she remembers anything about her so-called accident, she'll let slip why Ian came over here last night.'

Metallic clunks of a copper's hob-nailed boots striking slate tiles shattered the peace. Kirsten bit her lower lip, a habit if disaster struck. Her hushed sigh coincided with his firm grip on the hall phone as her husband strode towards the back door.

'Bjorn, is my mother in the kitchen? She's not on Abergeldie. I checked. It *is* urgent, otherwise I wouldn't worry you on the boy's birthday. A tragedy has occurred in Yass. A truck loaded with fertilizer has crashed into the MacDonald's home. The firemen saved the four littlens. Not their mother.' DI Raddick swallowed and tried to conceal his emotions. 'If my mum, Harriet or the girls get wind of this shocking event anything could happen.'

'Your mother is helping Helen with the cooking for tonight's barbie. Come in Paul. Jasmine and Rose have gone to collect Kendall's new saddlebag and a saddlecloth. Bart's altered it to fit Potchkin's sturdy withers. What a terrible disaster for those children.'

'Yeah. They lost their father last month in a logging accident up north. Tragedy always strikes in pairs, or so our big wigs reckon.'

'Paul, I want you to do me a whopping favour. With your credentials, the fire chief won't hesitate or argue. Fetch those orphans then bring them here. We've oodles of space and the two back bedrooms are comfortable. Your mum will love caring for the children, until the authorities stick their noses in to separate them. That will only occur over my dead arse. I'll give my legal bloke a ring. He'll be here later with his family. Surely, we can arrange something definite, before nightfall.'

'It sure will. I'll take mum with me to collect the kids. She'll be in her element nursing the youngest child. The kids can put a bit of huff-and-puff in those balloons. It'll keep their stressed minds occupied. Expect us shortly BJ, with a car full of sombre youngsters.'

'I'd like to catch the bum who stole Gigi's magnum of champagne, Paul. That prick will get a swift kick up the arse with my boot. Who's the hot chocolate for, Kirsty?'

'I'm taking this mug of cocoa to Gigi. She loves how Helen steeps it with honey from the native beehive, and a knob of pink ginger stirred with a cinnamon quill. This is heaven's ambrosia, nectar of the Gods, if sipped slowly. She'll sip this between dressing for the party and finishing her make-up.'

'Oh damn, there goes the phone. Answer it Kirsty on your way through. Paul's ready to leave and I'm going to ferret out the rat with thieving paws.'

1 pm: Bjorn rode over to Abergeldie's stables, only to find it devoid of human life. A cabin-trunk full of track gear sat alone amid a clump of grotty jodhpurs. Ian's battered saddlebag and a pair of his spurs were missing. From the stable's door hatch, he flagged Jaffa talking to Lanky Hancock. 'Hey there sport, have you seen Ian Ross anywhere.'

'No boss, not here,' Hancock's responded. 'An hour ago, I bumped into him in town. The stuff he selected must've cost him a fortune. It damn near blew me mind apart as he passed over five thousand quid for new furniture. I kept me mouth shut when he raised his fist in anger. Ross darted from the shop. The mongrel almost bowled me over in his hurry to reach his rusty crate. What's the bastard done now, boss?'

Bjorn's deep frown alerted both the men to keep everything under wraps. 'I've a fair idea. I'm off home. If you need me urgently, knock and walk through to Kirsty's office.'

Jaffa listening to them, pointed to the tongs in Bjorn's hand. 'Hey, aren't you going to season the barbeque? The possums have left their trademark on its metal cover.'

'Yeah, I better go clean both now. Ross can wait. Paul Raddick will be home soon with the MacDonald kids. I don't want them to go anywhere near that pit full of red-hot coals. Have you heard what caused the fire and their mother's death?'

'A catastrophe. I've warned both my terrors twice not to play with matches. A cigarette lighter may have caused the fire. The scientific blokes will investigate the ruin. Thank heavens the flames didn't spread to our property. My top paddocks are tinder dry. The tiniest airborne spark could create an inferno. And the home paddocks lack grass of any description. Stubbs of dried grass are barely visible amid cracks in the hard-sunbaked earth. Similar to Abergeldie's lower paddocks. Typical Ian, he'd argued when I mentioned that to him.' Jaffa then added, 'I cautioned him recently, if he dismissed Harriet and her daughter, I'd shred his cheque for the delivered haybales and make him swallow the pieces.'

Bjorn smirked then said, 'I can imagine what his face looked like after you warned him. It'd be interesting to see his reaction in your office. How did you get a gander at the total listed on his cheque book?'

'My secretary spoke and Ian turned. Then she saw the amount Ian had paid me for Abergeldie's last load. It was a spit in a sparrow's eye, compared to the total written on the last butt. I'd love to meet his backer. That person might enlighten the police about the dubious years he spent in Scotland. When he's plastered, I often think he may have Austrian or Swiss heritage. In my youth I visited Zurich and bought liverwurst sausages, sauerkraut and delicacies to supplement my meals, as all we paupers did in the war years.'

I've often wondered the same thing. Ian's a secretive bastard, and nobody has a clue whom he deals with overseas. He regularly receives Swiss mail and letters franked in Liechtenstein. Shit, the letters in Kirsty's desk. I forgot to give them to our postie.

Bjorn's thoughts changed to speech. 'Thanks for telling me about your wife. I keep my roadside drum locked in case strangers tamper with our mail. Yeah, I agree with you, Jaffa. Ian is a friggen liar and a thief. I could fill a book with his sexual indiscretions that would make that single red hair on your bald scalp curl. See you in the morning, sport. I'll tell Kirsty. She asked yesterday if I knew how your wife was after her breast biopsy.'

2 pm. A smirk lingered on Bjorn's face as his wife, with hands on her hips confronted him by the laundry door.

'Where is Kendall? He should be in the shower. His friends and our guests will be here in an hour. Harriet wants you to keep him occupied so he won't come anywhere near the kitchen.'

Bjorn nodded.

'Has Jarvis popped in since lunch? I gave him a bell before I went over to Abergeldie's stables. Oh shit. He could be in the harness room.'

Kirsten scowled. 'What is Stephen doing in there? Bjorn, have you been drinking?'

'You knew I meant Kendall. His new saddle and the horse rug are in there. Jasmine and Rose put them there on their arrival home from town. Gigi will be furious if he sees the saddle.'

'The saddlebag, it and the rug are in my office. Have a shower now. I need mine.'

'All right. Don't start nagging. I want to set up the portable bar first…'

'I won't ask you again, Bjorn. Your shirt stinks of sweat. Those black, greasy hands are disgusting. Our shower recess *is* vacant.'

The instant he stripped down to his undies, Kirsten nodded to Harriet in the hall. Quietly, her daughter and niece carried a cumbersome parcel from Kirsten's office to the kitchen and pushed it under the table. Harriet and her sister, Helen gasped with relief.

'This parcel will keep Kendall guessing. We'll take it out to the breezeway later.'

'Good idea. I'll fetch his other pressies now, Jasmine. Here comes his mother. Gigi doesn't know what you've hidden under that cellophane-covered box. Nor does Bjorn. Our secret.'

The girls giggled as they began setting the table with magical balls full of trinkets and fine streamers. 'Those darling little sprites will get a surprise. It's terrible to think of their mother dying in that horrible housefire. Their father died last year.' Rose sighed with tears welling in her eyes. 'I found this ragdoll in the airing cupboard. Poor thing, she looks ages old. A frock of ecru-lace hides her porcelain legs. They're jointed at the knees. Look, her cream satin petticoat is trimmed with delicate, pink-ribbon roses. She's beautiful.'

'The littlest girl, Margo will adore this doll. Shandra, her sister is five. Their brother Nathan is coming up seven. I feel deeply sorry for the poor kids, Rose.'

'So do I, Jassy. Oops, Jasmine. I forgot your mother will growl if I call you Jass.'

Ten minutes later Stephen Jarvis pulled in beside Paul Raddick's car, close to Jalna's breezeway. Alighting, he offered to carry the MacDonald boy indoors.

'I can manage Nathan, if you carry his young sister and help Laura, the eldest girl. The drug I injected into his bottom is beginning to take effect. My knee will act as a good door-closer.'

'Okay, I will Stephen. The police surgeon sanctioned your request over their blower. I intended to call him at noon. The youngsters will be fine here, in Bjorn's care.' Paul Raddick looked astounded as the eldest girl, Shandra jumped out of his car and scampered to grab a sandwich, before anyone knew both their vehicles had arrived.

Margo snuggled down in his arms. She whimpered as he removed their black, soot-doused case from the rear seat. Shutting the car door with his knee, Paul carried the case and the baby girl around to where his mother, Harriet was standing by the first cool room. Shandra clung to her skirt. The child's huge smoke-inflamed eyes, gazed up at her with a pleading look.

'Mummy can't come home to us, can she? The firemen couldn't save her or our pup Mopsey. Are they both in heaven with our Daddy, Mrs Raddick?'

'Yes darling. God's angels have wrapped their wings around your mummy. Come with me to the kitchen. Nathan's asleep, so we best be quiet and not disturb him. Mrs Svensson is bathing your little sister in a warm bath. Would you like a glass of iced lemonade, Shandra? Aunt Helen has taken a jug up to Kendall's mum. Sit on this chair sweetie, and I'll tell you a real secret.'

'Ooh, I *love* secrets. What is it Mrs Graham.'

'Call me Aunt Harriet. Do you remember Rose and her cousin Jasmine?'

'Yes, I do Aunt Harriet. Your daughter taught me at Sunday School last year.'

'She and her cousin have hidden Kendall's new saddle under this table. Perhaps the girls might let you help them carry it out to the breezeway. We'll ask them in a minute.'

4 pm. Kendall confronted Kirsty and his mother in the hall.

'The party is off. I cancelled it, Mum. How can I celebrate my birthday, knowing those traumatised kids are grieving for their mother? Keith just told me he saw the kids in her car as he drove past the house. Apparently, their mother hurried back to disconnect the power and to fetch her keys. An electrical fault may have caused the fire, which engulfed the front hall and she couldn't escape in time. Or that's the theory.'

5 pm. On conclusion of the day and before dinner, Bjorn asked everyone to gather around the back table while Kendall cut his cake. There were no cheers or accolades.

Harriet's index finger lined up with a solitary gift on the table.

'Excuse me, if I may say a word? Kendall, would you help this wee lassie remove the cellophane wrapper off that box. Then we *can* celebrate your birthday. Your other gifts are unopened in your old room.' She winked at Gigi, whose dimpled-smile grew larger with every breath.

Excited gasps echoed through the breezeway. Bjorn's mouth sat agape.

'The missing magnum. Well, it's not what I suspected. Where's the saddle?'

Shandra squatted and with the biggest grin pointed her little pinkie and whispered, 'Look under the table, Kendall. My short arms can't reach that big wrapped parcel.'

'Your goblets are full, courtesy of Bjorn. Please join us in toasting my son's success. As you all know now, the party has been postponed until next weekend…'

'Mum, I was going to tell you this, before our little guests arrived. I have been granted a full scholarship at Sydney University to study medicine. Tomorrow Keith and I will be driven by his mother down to Canberra. Our flight from there to Kingsford Smith Airport leaves at eleven-fifteen. Sorry to disillusion you, no party Mother dear.'

'Deceive me, you mean. Please charge your glasses to Kendall and his future at university. One day soon, he might return the favour and charge you all a whopping fee. Now your speech, please son?'

'Give me time, Mum. Thank you everyone for your gifts and good wishes. My sincere wish is for Shandra and her siblings to find contentment with Bjorn and Kirsty. Or until their aunt can fly out to Australia from Ireland.' He bent down and kissed the child's flushed cheeks. 'We all realise this may not occur for some months.'

Gigi gasped. 'I ought to scold you Kendall, but I'll let you off this once.'

'Now Mother dear, don't look startled. I am not leaving home for good. I'll be in contact with you either by phone or a letter, as time and my studies permit. Buzz Keith's mother if you can't reach me at my digs in Camperdown. Kay will know…'

'What mischief have you and Keith devised now?'

'I promise to ring you well before our plane leaves at eleven, okay. Now I confess. I forgot to show you the telegram I received on Friday from Sydney. Keith has accepted his scholarship in Veterinary Science at the same University.'

'Who's given you this grant, Kendall?' barked his father, walking in unannounced. 'You can't expect your mother, or me to shell out thousands to keep you in a fashionable squat in Sydney.'

Doctor Jarvis coughed. 'Let me handle this, Kendall.' His anger and savage glower targeted Ian Ross. 'A full scholarship means nobody will pay a cent for Kendall to study medicine. I have given him my blessing, plus current books to help him with endless lectures. If he needs financial backers, Bjorn and I will be pleased to cover his expenses. His accommodation will also be catered for, Ian. There's nothing you can do to help, or hinder his progress.' Stephen deliberately omitted saying "your son".

About to walk away, Ian Ross scowled at his son and sniggered in a low drone, 'Why should I care a damn? He can do what he wants, as long as he doesn't come snivelling to me for money. I'm going home, Gigi. You can stay and listen to their crap and then some fool can drive you over to Abergeldie.'

Bjorn waited until Ian's car pulled up the drive before he approached Kendall. 'That frown of yours I couldn't ignore. What's worrying you, matie?'

'That prig is, Bjorn. He threatened to shoot Potchkin the last time I rode him over here. Can I leave him stabled with your horses here on Jalna? It will ease my mind. Mum says she'll ride Wallaby Downs over tomorrow, before I leave with the Brown family. I tried to talk her out of it, but she wouldn't listen to me, as usual.'

'Need you ask? Potchkin will be attended to by my vet. Kendall, your mother is doing a terrific job of raising cattle *and* sheep. In a few years she will be the boss-cocky of Abergeldie. Ha, then your old man will be forced to accept her orders.'

'That'll be the day. I can't imagine it happening. Thanks for your financial support and persuading me to sit for my original, medical science exam. Can you imagine what his reply would be, if he knew that you and Jarvis sponsored my application to the university. It'd send his mind into a whirlpool of frustration and wipe him off the map. Do you know that he keeps a loaded rifle under his bed? I went in the den yesterday to answer his phone and saw that damn tan briefcase I hate, almost as much as I do him.'

'Kendall, if he ever threatens to shoot your mother, I'll have him in court quicker than he can blink. Harriet Graham will take good care of your mum, while you're studying twelve hours a day in the big smoke. Here's another three grand. It should tide you over until you find suitable digs.'

'Really Bjorn, I can't, no I won't need it.'

'Keep the money. I know damn well how hard it is to manage on a pittance. Slaving in damn rock quarries almost broke my back and dulled my mind in those bitter Italian winters. We battled to snatch a meal of horseflesh. War created its own hazards without the Italians bashing us senseless with their rifle butts. My frostbitten toes worry me at times, as you know. They locked our blokes in three-foot-high cement cells we called the hole, for twenty-four to forty-eight hours in one hit for the slightest misdemeanour. I stole a chunk of gouda-cheese from the quartermaster's hut. It tasted beaut to us dehydrated, ill and undernourished pilots. Flying planes in those days was sheer hell, with flack pelting our wings on strafing runs over Norway, Austria and Germany. I couldn't get back home quick enough, Kendall.'

'I'm an ardent reader of that period of the war in Europe. It must've been utter hell slaving your guts out under such harsh conditions in mid-winter. We don't realise how lucky we are working in this sweltering heat on the land, Bjorn.'

He just nodded.

19

6am Monday morning. A fox howling in Jalna's lower paddocks caused Kendall to hop smartly out of bed. He took a peep at the hot breakfast on the covered tray.

'Ooh boy, am I hungry! I better rinse the sweat from my body and shave before I tuck into breakfast. No. I'll eat first and then shower. Mum promised to be here by seven and she'll get a fright if she catches me in the nuddy.'

At least the weather looks promising. Now, what will I wear today? Casual clothes will be best to travel in. Planes can sometimes be stuffy until take-off. Kendall turned his nose up at the clothes on his bed.

'This pale green shirt goes well with my brown trousers.' Fumbling fingers rummaged through the wardrobe rack. Undecided, he held up a sage-green tie with fine-swirls of dove-grey. 'This one's perfect. It suits my complexion. Shoes brown or black? I've packed my brown tweed coat and rain mac. Toiletries are in, so are three belts. Better pack this fawn one. It belongs on my tan slacks. Underclothes are folded on top of my summer dressing robe with an umbrella in this side-pocket. My shaving gear and everything I need is tucked in there. All done.'

Kendall locked the suitcase and smiled. 'Now I better see if Mum's arrived. I didn't hear the car. No, she's jockeying over on Wallaby Downs. He's a magnificent stallion. I hope she leaves him here with my horse. Otherwise he could be a target for that maniac's bullet. The idiot over home shoots everything that moves, especially if he's in a drunken mood. It beggars belief why she doesn't get a divorce. Jarvis's records prove the extent to which that creep has brutalised my mother. What more proof do they need of his bestiality?'

'Your mother's tethering Wallaby Downs to the back verandah post, if you're worried Kendall. She'll be in soon, matie. Why the frown?'

'I feel guilty for being deceitful, Bjorn. I should've told her days ago that I was leaving today. A stupid mistake, a neglectful oversight.'

'She understood. Hey, there's a comical side to this dilemma. Ian walked in at that precise moment and heard you telling everyone about your scholarship. The startled look on his mooch I captured on film. I'll post you a print later. Whenever you feel miserable take a gander at the photo. It will brighten your day. Is this all your luggage, Kendall? If so, I'll carry it out to the front verandah. The Brown's car is turning in our drive now.'

'I appreciate you offering to take care of my horse, BJ. It'll be a huge relief off my mind while I'm listening to medical lectures and taking notes at university.'

'Potchkin will be safe in my stables. One of the muckers will trial him every morning. He loves a sand roll after each run. The girls are anxious to give you a farewell hug, they're in the hall. Everyone else is on the front verandah, ready to wave goodbye.'

Bjorn glanced at Gigi. She looked depressed on seeing her son leave, perhaps for the last time in months. As the Brown's car cruised back up Jalna's drive, tears in her eyes flowed. He knew the anguish was building to crescendo. A mother's love for her son's parting, now under a cloud of misery, caused Gigi to turn away. The turmoil in her traumatised mind made it impossible for her to think of a future without Kendall and his support to keep her sanity against an insufferable foe. Her husband.

Please God, keep my son safe in whatever he undertakes to do in Sydney. I hope he finds it in his heart to forgive his father's anger to me. Kendall deserves to lead his life in peaceful surroundings. I have caused him endless misery until now, because of my insane fear of Ian. In time, Ian may change and see the folly of an impetuous temper. Her silent prayer ceased as she whispered to the wind. 'Goodbye my son, until we meet again. Perhaps in a home where peace reigns. And sadness is forgotten.'

6 pm. Kirsty suspected her husband knew more than he'd let on about their ex-employee. She recalled the anger reflected in Bjorn's eyes.

'I think Ian was beastly to Kendall over the university grant. Yet he wasn't prepared to put a damn hand in his pocket to help the kid with his finances. I can't think of the last time I saw my husband incensed to that degree. I'm glad Ian's over on Abergeldie. From there, he can't bash Gigi senseless. She was scared of what he threatened to do to Kendall, if the boy disobeyed him by staying here on Jalna this weekend. He's a good lad and he idolises his mother.'

'Bjorn will shoot Ian if he harms her from tonight on. He'll shoot to maim, not kill. I'll think of something to keep them apart, for her sake. Gigi craves love, not aggression. Although I was astounded when she wanted to go home this morning. Ooh, how I'd love to drop a handful of sleeping capsules in that beastly man's wine. Then his control over her and the freedom she deserves will be non-existent. No, the pills were a foolish idea. A hard whack across his mouth would be better. I hope Ian doesn't barge in and catch Bjorn on the phone to their solicitor querying of his last tax refund. That would cause pandemonium and upset everyone in this peaceful, countrified town of Yass.'

Ian Ross never returned to Abergeldie all that night. Neither hide nor hair was seen of him in or around town. It was as though he'd walked off a cliff into oblivion. Bjorn had spoken to local farmers and came to the same conclusion that Ian had booked into a hotel either in Goulburn or in Canberra.

'Stephen Jarvis popped in there an hour ago to give Gigi her injection. He may drop by again this afternoon. I have an idea which might work, Kirsty. Your equestrian coach is visiting Abergeldie tomorrow to collect his saddle, I'll ask him to speak with Gigi. Jack offered some time ago to mediate between the couple. She trusts him to keep a promise, as he does her. A brief discussion may save a lot more conflict in that family. His idea is to persuade her to join their equestrian team, especially with a challenging team on the horizon.'

'Gigi is an excellent rider. You know that Bjorn. And so is Kendall.'

'Okay, you can tell her what I suggested. I'll be working with the men, re-fencing our top paddock if you need me. Your handset is on the hall table. My two-way speaker's in the jeep, with waxed cartons of frozen fruit juice and ice.'

'Have you put both the eskies in there, Bjorn?'

'Yes, I have. Everything's under a tarpaulin to keep cool. I shoved a couple of four-gallon drums of fresh water in there at six. I better be off, or Lanky will be wondering where I am.' Ready to leave the office Bjorn turned. 'Kirsty, if Ian rolls in here drunk don't let him in the house. If he hasn't put in an appearance here by four, I'll ride over to Abergeldie, after Lanky and I unpack the jeep. Harriet knows I feel like scattering Ian's brains after last night's debacle.'

12 noon. Bjorn buzzed the homestead on his two-way radio. 'Have you heard how Gigi is this morning? Harriet rang me to say, she may've suffered a fall? I'm pleased he's there. Stephen knows what to do if Lord Muck threatens more violence. I'm glad Kendall's not home, Kirsty. He'd have throttled his father. With the boy in Sydney, it means Ian can't threaten or bully the boy. This break will allow his mother to recuperate after her last *suspected* accident.'

The house remained quiet with its occupants trying to cope under difficult circumstances. Doctor Jarvis arrived to find Gigi lying in a pool of blood on her bedroom floor and assisted her to the bed.

'Don't try to move, Gigi. The contusions on your knees and ankles will take time to heal. Until the swelling subsides your legs and bruised arm can't be x-rayed. If you don't mind me asking, what caused your fall? I need to know the truth to write it on your chart and medical history. You'll feel a little faint after this injection. The effects will wear off in an hour.'

'I climbed to get my hatbox and toppled off the embroidery stool. I thought I'd broken my left wrist, until it moved flexibly. This bump on my head must've happened in the fall. I remember hitting the wardrobe. Then everything went black, Doctor Jarvis.'

'You should rest for a day or so. Have you or Kirsty heard anything of Ian? I saw him yesterday in town. None of the locals know where he is, or where he might've gone.'

'No, I haven seen or heard from Ian. He'll turn up when he wants something. Ian refuses to confide in me. It's sad really. We're renovating Abergeldie and the painters are having a break on the back verandah. I did trip over their sheet protecting the hall floor about an hour before I fell off my dressing table stool. Clumsy me. It was an accident. Ian's not responsible. Carrying a tray, I wasn't concentrating and didn't see the mat's upturned corner. I'll be fine tomorrow.' Gigi clasped her throbbing temples. 'I tried to save myself from falling and that's all I remember.'

Replacing his surgical instruments in his media bag Jarvis nodded. 'You have slight concussion, Gigi. The swelling above your left temple will subside in a day or so. Accidents are unpredictable. The cut on your left elbow isn't serious and didn't need suturing. Try not to bump it or wet the dressing. I'll pop in tonight if your headache hasn't eased. Take two of these capsules an hour before meals. They'll ease both your nausea and the pain.'

'I'm a bit anxious. Ian never strays far from the house, especially if painters are finishing the walls and doors. Not unless he's a little under the weather. Lately he's more interested in grape-growing to brew his own wine. So far it's textbook stuff.'

'Well, I can contact the police. I do know Ian drinks to excess. Most drunken people don't hurt themselves if they fall. He'll be alright. Don't worry needlessly, Mrs Ross.'

'Please, I hate missus. It sounds harsh. Call me Gigi, it's not so formal, Doctor.'

'Harriet is going to bathe your knees with warm, salty water. It's an antiseptic and it will sting ...'

Abergeldie's cook interrupted Jarvis to say that her daughter Rose and her niece Jasmine had seen a man similar to Ian Ross in the barber's shop in town this morning. By her grim expression Stephen knew something drastic must've occurred. 'What's happened? Harriet, is one of Bjorn's men injured?'

'Yes, Doctor and I've rung the ambulance. Ross Meadows has sliced his leg open above the knee. The bowsaw slipped. Bjorn tied a tourniquet on his lower thigh. It has eased the flow to a degree. They were re-fencing close to "Wave Rock", so the newborns can't fall down its deep crevices. The ewes would try to save their lambs and also get injured themselves. Although the rock is beautiful, it is a really treacherous area. A stray calf died up there last summer while trying to reach a pool of water caught in a crevice.'

'Right. I'll head straight up there, through the top paddock. My car runs well over rough, dust-pocked ground. I'll keep away from the rough patches. Tell Kirsty I'm on my way. It may take me ten minutes to steer clear of that rocky outcrop.'

Hearing two shots in quick succession, Harriet dropped the phone. 'That's Bjorn's urgent signal. If you follow the backtracker in your car, he'll know the exact spot where they're working. Jamie should be here now.' She heard a cooee from the back door. 'We're coming, Jaime. Lad, show Doctor Jarvis where the men are located. The ambulance should be pulling in through Jalna's gates soon. Stay with them. Bjorn might need your help to lift Mr Meadows. You know how to use a walkie-talkie.'

'Sure do, Missus. Come on Doc. I'll give ya a lift on me horse. It'll be quicker than your old crate.'

Stephen Jarvis sniggered, 'Old crate indeed! It's a damn new vehicle.'

Harriet laughed and passed Jarvis his medical bag that he'd left on the car bonnet. 'There's a fully equipped surgical kit in the jeep. Bjorn keeps it stocked with new bandages, sterile instruments, pads, scissors etcetera if you need them, Doctor.'

Jarvis looked over his bifocals and grinned. His hands clung to Jaime's ribs as the horse galloped through the home paddocks. Stirring up a flurry of dirt and grit she jumped a dry creek bed that shook the hell out Steven's body. The mare then landed on all four hooves, softer than a newborn's bleat.

Glad this ordeal was over Stephen hopped down from its hindquarters and landed with a sudden thump. Balanced, he massaged his sore, stiff bone-shaken rump then headed over to where Ross Meadows was lying in the shade of scribbly-gum.

2 pm. 'You did a good job of clamping those capillaries Doc, a temporary measure. We're ready to roll. Base will contact his wife to let her know where we've delivered Meadows. As of now it'll be Yass Private Hospital. The surgeon on call will take over from us. He'll supervise him being prepped for theatre. The sawblade missed his left femoral artery by a millimetre. After surgery and rehab, he'll be hobbling on crutches for some weeks. We'll send you a copy of our report for your files tomorrow, Doc.'

'I'll be in Yass later today to see a patient, and I'll speak to the surgeon. Better still, I'll collect his wife. Their farm's only a hundred yards downwind from mine.'

4 pm: Bjorn's men secured the last rust-proofed wire-clip in the fence post then stacked their gear in his jeep. Unable to think of anything but the accident, he let Arthur Hancock drive through the lower paddocks.

'Unload the eskies and those rotted fence joists first, Lanky. I'm absolutely strapped. It's been a hell of a day. We can't afford another accident to happen to one of our blokes. I found it gut-wrenching to watch Meadows in agony. None of us were capable of helping him or to ease his pain.'

'The shot Jarvis gave him in the arse worked quick. I'll buzz the hospital tonight ta ask how he's fairing, BJ. Meadow's looked pretty crook before the ambos' shoved him in their bone-rattling jalopy.'

'Don't phone, Lanky. Jarvis will be in town all afternoon. He'll give me a buzz after he's spoken to the surgeon on duty. He promised to let me know if there's positive news. Pass my message to the other blokes. Phew it's a stinker. A brisk rub under a cold shower will cool my grungy armpits and thighs.'

Bjorn slung his sweat-matted Akubra on Jalna's back table and sauntered through to the kitchen. 'What a shit of a day it's been, Helen. Where's my wife?'

'She's talking to Gigi in her room. Their afternoon tea is here. Dinner won't be ready until six-thirty or seven. The oven flue was clogged. It smoked something awful until I fiddled with the gizmo … err lever. It took ages to clear.'

'Remind me to ring the chimney bloke first thing. His boy will fix it. I'm off to soak under a long, cool shower. Fetch a midi from the small camping fridge Helen. I could down a keg of ice-cold beer, I'm so damn thirsty.'

Still wearing the sludge-drenched singlet, Lanky Hancock joined his boss on Jalna's back verandah to discuss the day's events.

'The Doc arrived just in time this arvo. Meadows could've lost his leg.'

'You're wrong there, Lanky. The ambos are trained to handle all kinds of injuries in an emergency. Jarvis works very hard with our crew if something goes wrong. Phew, it's stifling. It should cool down, once the Snowy wind rides in around dusk. This oppressive dry heat is severe enough to fry an egg and shrivel the soles of a bloke's boots.'

'Thanks for the beer, Boss. I'm off ta take a shower before the dinner gong blasts the peace. Hope its cooler tomorra. This sweltering summer heat can't last much longer.' He brushed the flies away with a sweat-battered slouchy and headed to his bunkhouse.

Lanky assumed Gigi would be going home in the morning. His loyalty lay between Bjorn and Abergeldie's new manager, Tom Marden. Bjorn had released Tom from his contract to work on their property, knowing Tom would keep Ian under surveillance. The same principle had applied to Mrs Graham. They both disliked and mistrusted Ian Ross for obvious reasons. Harriet wouldn't tolerate his binge drinking, not while she was in charge of the kitchen and supervised their household staff.

20

7pm. In the stillness of night, a loud whisper would echo more than a mile away. However, he knew what must be done. Bjorn heard a familiar voice roaring at his wife at the front door and headed to confront her abuser. Pointing his rifle at the intruder, he signalled Kirsten to leave. A bleak wind drowned the howl of foxes as he moved up the hall.

'What are you doing here, Ian? My wife didn't invite you to dinner.'

'I demand to speak to *my* wife. If she doesn't come with me, I'll drag the bitch home by her hair. She's not sleeping under your damn roof another night.'

'We're entertaining guests, come back tomorrow morning. Then we can talk when you're sober. If you don't leave my property now, I'll call the police.'

'You needn't bother. I'm leaving Yass tonight. Tell my wife, I'll come home when it suits me, not her. I've had a gutful of her criticising everything I do. And that goes for you, Svensson. You can all go to hell. I need peace, not argumentative grouches snarling at me.'

Kirsten's swear words were mild compared to the men's, as she went to speak with her music tutor.

'Stay inside, trust me Ted. We've got to make sure nobody leaves the lounge room. Bjorn will tell you why in a minute. Drinks are on the house.'

'What if I say we're celebrating the good news? I did hear Gigi is thinking of taking over the management of Abergeldie.'

'She keeps the bookwork and funds under control, to protect their finances.'

'In my estimation she'll succeed. She's very determined and a talented woman. Where's Ian? I know of late he's taken himself off to Canberra and Sydney without mentioning a thing to his wife, or Tom Marden. I have a feeling that he's been visiting illicit dens of iniquity. He said something the other day about the young women there who were incredibly seductive, some were beautiful. They must've catered to his taste of exotica. I thought it odd at the time. Gigi can't afford the massive loss of their wool clip price. Most of our neighbours will club together to do the shearing, crutching and dipping of their mob while Ian's away enjoying the spoils of *her* hard-earned money.'

'Exactly what Bjorn said this morning. His men will round up the casual shearers, and they're going over there tomorrow at six. What he has in mind might work, Ted.'

'Going to the lower fences and let their mob run with yours, aye Kirsten?'

'Yes. Arthur Hancock set more traps up top yesterday. Tonight, he penned the ewes and their newborns in with a flock of wild geese. Our four kelpies are guarding the main herd. Those dashed foxes won't get another chance to kill more lambs.'

'I know how your husband feels. It's heartbreaking to shoot dying newborns and the mums with their entrails flyblown.'

Kirsten didn't reply to Ted. She walked on ahead and announced to her guests, 'Well, you lot look miserable. How about a tune on the piano? Ted's promised to act as host while I play a tune. Gigi, a song? Your voice is clearer than a lark's …'

'Not at present,' she interjected, 'I don't care to sing now. Later perhaps, Kirsty.'

This over-polite affirmation came as a shock to Kirsten. *What's wrong with Gigi? Usually she's the liveliest guest at our dinner parties. Ian's absence, of course. In his unstable frame of mind, he'll be no loss. I wonder where he is. She objected to my suggestion. Has he left her and gone meandering? I know he's a womaniser and it is possible. He doesn't give a damn about her, or Abergeldie. Why! Tom Marden might know what Ian has planned.* On reflection Kirsten frowned. *Gigi is handling the financial accounts of Abergeldie well. Kendall will take over when he comes home in September. Our men and neighbours will make sure she's okay financially. Bjorn and I will be her backstay if Ian decides to leave home for good. He's threatened to do it dozens of times.*

Gigi sounded weird on speaking for the second time within seconds. 'Why don't you play 'Ballad for Adeline' and sing Kirsty? My throat is sore.'

'We understand. You do look tired Gigi.' Ted Hardacre sensed all wasn't well in the Ross camp and took over. 'Why don't you go and rest a while? None of your friends will leave while the whiskey's flowing and the beer's on tap. Let's celebrate your good fortune. I must say Gigi, you've weathered the storms of sheep husbandry. Now it seems you should have a little spare time to celebrate your success in raising a huge mob.' He embraced her gracious smile. Sensibly he refrained from ridiculing her husband over his insistence to leave home. Ted boosted her morale by giving her a hug.

'Anyone care for a drop of this delectable nectar?' He pointed to the tantalus, while trying to keep a peaceful ambience. He thought better of this idea and revelled in being the host. With a napkin elegantly slung over one arm, he began pouring the drinks.

The women's hands wavered. They were content with tea. Kirsten reflected on her choice. *It'll look suspicious if I don't toast with something more appropriate than tea, especially after suggesting it.*

'A nip of whiskey, no ice for me. A weenie one thanks, Ted.' Her fingers measured a dram. 'Here,' she passed him a cut-glass goblet from the bar fridge, 'half-fill this, please. I have accounts to finish and can't afford to have a befuddled brain.' Her lips twisted and her head wobbled. Everyone laughed. This comical performance created the ambience for the guests to wallow in her humorous repertoire.

After several rounds of drinks, the front door slammed. Yet nobody walked down the hall and Kirsten delivered her prepared speech. 'Bjorn should be in here soon. Now I'll ask our special guest, Ted Hardacre to present the toast to Mrs Ross.'

'I agree with Kirsten, you deserve a medal for persevering under stress. Gigi, you my dear, have excelled where others have failed. It isn't the easiest of jobs managing a large homestead and a massive staff. I speak from experience. Look forward to a contented life with your son. Don't look back. This is the era of a bountiful harvest of sorghum, sunflowers and fields of golden grain. Love comes from reaping the rewards of harvesting and hard work. The running of sheep is a secondary consideration, where you and Kendall are concerned. Remember dear, as your friends, we will be honoured to help with the wool-scouring and working the harvesters.'

Playing for time Kirsten took over to say, 'I'm going to tell my husband to hurry. I'll be back shortly.'

Before she could leave, Gigi stood up and her eyes drooped. Liquid spilt and splashed everywhere. Helen Raddick used a dishcloth to wipe a mess off the floor. Gigi glared at her frock, now tea-stained. 'I'm sorry, Kirsten. It was an accident. I am a bit tired. Will everyone excuse me, if I toddle off to bed?'

10 pm. Kirsten caught her husband stepping from the shower. Shutting the bathroom door, she asked eagerly. 'Darls, did Ian tackle you? Your clothes smell like a brewery.'

'I stunk worse than a brewery. He was his usual repugnant self. Gigi will go home to a peaceful house tomorrow. Ian's flown the coup. He told me to mind my own business, so I told him where to stick his head. Where he sits on it. The bastard is heading to Melbourne or some such place. When and if he does return home, he *will* answer to me. Lanky has just about had a gutful of him, and so has Tom Marden, their manager. Kirsten, he will make Ian toe the line. He will knuckle under with the weight of his men threatening to castrate his balls. He deserves to be strung up and four-quartered, like they did to criminals in the eighteenth century. In the olden days, pirates roamed London town with swords drawn ready to splice all evil-doers apart.'

'You're making my blood run cold. Bjorn, stop this nonsense now. I agree Ian does deserve some form of punishment, because of what he did to Gigi. If you and Jamie hadn't found her unconscious and injured under those bushes, in Abergeldie's top paddock, she would've died that night.'

'What annoys me most is why Ian lied to Paul Raddick and the police. Ian then reckoned he'd never driven their Landrover all day. Jamie and I had both touched its bonnet before mounting our horses. He also denied Gigi had been missing since five that evening.'

'Yes, I remember how you blew into your burnt palms to cool them. It's nothing short of a miracle that she recovered in hospital a month ago. We should have phoned Kendall the next morning.'

'Why worry him unnecessarily! What could he have done in Sydney? What a daft idea, Kirsten. If we don't tell him soon, either Paul or the local copper's will. And we can't afford to let that happen, not while he's studying to gain honours in these exams.'

21

5am. Half a bottle of wine, plus an uncorked one welcomed Gigi in the breakfast room. 'It looks as if Ian's drunk himself stupid overnight.' She searched the house and found a note in his den, which she quoted, "My mother is ailing. I'm leaving now for Edinburgh. I could be gone for ages. Ian." Mystified, she thought, *Why the dickens didn't he phone me over on Jalna last night. Oh well, it makes me more determined to work the sheep and homestead on my own. My first job will be to clean his office. The room needs tidying and a good dust. At least he won't be growling, or constantly demanding his meals on time. His mother must be really ill, or dying for him to leave suddenly. He's hedged and refused to tell me anything for ages regarding her health.*

5.39am on Jalna. Bjorn towelled himself dry while discussing Ian with his wife. 'There's an uncorked bottle in your wardrobe. Don't say it Kirsty. I'm not in the mood for joking.'

'Where is he now? I heard him roaring at you on the verandah last night.' Kirsten's furrowed brow narrowed and it showed her concern for Gigi.

'He threw the bottle at me and called me a liar. Madder than a hatter, I told him to stick it up his...Well what he sits on. The bastard threatened to kill me then shot through at night. And good riddance,' Bjorn retorted, drying his hair. 'I felt like booting him down the front steps. He rode over, that's why we didn't hear the roar of a car engine. The silent approach works every time. He's no loss. I feel sorry for Gigi.' Bjorn wore his underpants on his head, complements of his wife.

'Why? Now she's free to run Abergeldie without his interference. In time he may return home. Bjorn, you often say a rotten penny flips the right side up. Oh dash, there's the front door bell. I'll go while you finish dressing.'

'Good morning Ted. Come on through to the kitchen. Care for a cup of coffee or tea?'

'Love a cupper. Tea will be fine thanks Kirsten. I just popped in to ask Bjorn to co-sign this lease for my farm. It *is* urgent. With all the mystery and skullduggery last night, I forgot.' Ted looked a bit confused. Kirsten sat opposite her choirmaster as Bjorn entered the kitchen.

'Ian's gone meandering, if you know what I mean, Ted. We were compelled to come up with something fast. Gigi's now free to move on with her life without a nagging husband to bully her all the time. She is feeling despondent, because Kendall won't be home for his birthday or Christmas, may be well into next year.'

6 am. Checking through the previous day's mail, Bjorn picked up a Swiss-franked letter on his wife's desk. This looks interesting. It's another letter addressed to Kleinhardt, marked attention Ross written on its corner. *Kirsten will be bitterly mistaken if she thinks there might be the slightest bit of decency in that Scottish bastard. I'm damned sure I can't see any descent traits in Ian.* A compassionate man, Bjorn didn't believe in violence, unless provocation warranted it. *An idea unabsorbed by a fragile mind, Ian refuted my suggestion for him to leave his wife.* Unconvinced it would improve either of their lives, Bjorn's suggestion had failed to penetrate his sozzled brain. Ian had purposely rejected every proposal put forward by his neighbour, and reckoned they lacked commonsense.

Bjorn re-read Ian's signed deposition in which he had transferred Abergeldie and all its landholdings to Gigi.

'I better put this document and those letters in Kirsten's desk in our bank for safekeeping. Someday they may prove how the swine has dishonoured his marriage vows.' With the receiver poised to ring Paul Raddick, Bjorn hesitated.

Instead he dialled Canberra and asked to speak with Senator Bucknell, a close friend.

'Hi there, Peter. Do you remember the letters I spoke to you about regarding a certain neighbour of mine? What do you suggest I should do with them? Yes, I'm listening. Okay, I'll leave them in this drawer and lock it. Kirsten keeps the key on a chain around her neck if she's away, or if she leaves the office. They'll be safe there. This line's a bit hazy. You did say you'll be home this weekend? We're to expect to see you on Friday, bye.'

Bjorn bundled Abergeldie's deeds and the letters together in a rubber band and dropped it in the drawer.

'Thank goodness that's done. I'm pleased we won't be summoned for withholding official mail. The Post Master General takes a dim view of mail tampering. I mustn't forget to tell Kirsty what Peter Bucknell recommended. I'd love to know who M. Kleinhardt *is*.' Bjorn flicked the just received envelope over and read its return address. 'The writer could be a woman? The same script on all the other letters is different. This one looks as if was inscribed by a firm, masculine hand. Time will unmask this mystery.'

Bjorn closed the office door before confronting his wife in the kitchen.

'There a letter I just locked in your desk's lower drawer with the others, Kirsten. It looks as if we better add another guest to Kendall's party. Peter Bucknell will be here first thing on Friday. He and his secretary send you their regards. Peter may bring Dee with him this break. He says it'll be good to be home for a short spell, Parliament's in recess for the festive season.'

Bjorn turned and spoke to their cook. 'Helen, make sure there's ample vegies and fruit on this weekend's menu. Senator Bucknell's a vegetarian, he prefers fresh fruit to stodgy rice pudding, or hot sweets.'

'None of my puddings are stodgy. The Christmas plum puddings and cakes must be moist. If they're undercooked I turn them into fruit balls rolled in coconut, Mr S.'

Reviewing his notes, Bjorn reminded his wife of what Ian had written on the property transfer.

'Did I tell you, all the money listed in their joint account he signed over to his wife? Gigi is the sole custodian of their estate. And it's not before time. I threatened Ian by literally twisting his ear and he soon caved into my demands. He rebelled and reckoned Abergeldie still belonged to him. As you know, I refused to be bullied by him.'

'Bjorn, that's water under the bridge. I'm going to see if there's more Christmas cards and surface mail. Bills we don't need, not at this stage, with a party underway. Be back in ten. I'll ride Sun Bronze up to the main gates. Have you anything to post, Harriet. and I'll leave it with my letters in our RMB. The postman will collect the letters on his return run around four.'

Mrs Raddick pulled a clump of letters from her apron pocket. 'These envelopes have Jalna's address written on the reverse side, thanks Kirsten. With luck, they should be delivered in the UK and Switzerland within a week.'

22

1963. Time travelled down the years until late October and in her silent sphere of peace, Gigi focused on unravelling a host of unpaid accounts, ones her husband had accrued on his overseas jaunts. *This problem is a challenge to me. My son is due home today, and I'm following Stephen's advice by ignoring Ian and his grumpy moods.*

7 am. With tears mounting in her eyes, the black-edge letter and an envelope fell from Gigi's fingers to her lap. Pitifully sobbing, she pulled a tissue from the box.

'This horrible news of her death in Switzerland is the last thing I need today. Why didn't Ruffina's husband phone me? Maurice knows Abergeldie's number. Apparently, he hasn't changed. Given the chance, he would argue with a dead dog to satisfy his narcissistic nature. I detested his pomposity and he treated Ruffy far worse than Ian treats me.'

What's the use of sitting here feeling miserable? I'll have another shower and slip on fresh undies and a cool blouse. These soft-green cotton slacks will be perfect to wear in this sweltering heat. The problem is my hair. A French-knot will be best and it'll be far cooler than this saturated mop. "Sensual-Passion", this gentle-pink lipstick feels smoother on my lips than the rich coral. Now shoes, my mid-grey sandals are comfortable and their soft leather won't rub my heels.

The breakfast tantalised her tastebuds. 'Your mango and pawpaw pancakes look delicious Harriet, topped with passionfruit and maple syrup. No, I don't feel hungry.' Scrunching the letter full of rotten news in her hand, she continued to mutter, 'Normally I love munching on your crisp bacon. No eggs this morning thank you dear.'

In between reading the morning paper she sipped her tea.

'It'll be a pleasure to eat what I like without being told what I can and can't do. Ian might sleep for hours in his den. He can be boring, if deprived of a liquid breakfast. My worst fear is will he and Kendall argue? Did you hear him come late last night, Harriet?'

'No Gigi. Rose took a pile of clean linen in the den at six and found him asleep on his bed. She kicked his briefcase and cursed it. There aren't any flight tags on his luggage.'

10 am. Gigi selected a tight rosebud from her white Dolly Varden basket and tested the water. It came mid-way up the glass insert. She sniffed the apricot roses, their tight buds and lemon verbena amid sprigs of white Gypsophelia, or Baby's Breath.

'Shall I put this arrangement on the dining room table, Gigi? I saw you cutting those delicate buds with your secateurs at six. A welcome home to Kendall?'

'Yes Harriet. My son adores roses. Like you and the girls, he thinks all flowers are God's gift to brighten our miserable lives. Did the phone ring while I showered?'

They both jumped as the hall phone buzzed.

'It's probably Kendall. I'll be careful carrying your Dolly Varden basket to the breezeway. The roses won't wilt in this breathless heat and its cooler out there than in the house.'

'Good morning, Mother. You won't mind if I bring two friends home for breakfast? My hungry car is guzzling petrol from the bowser.'

'Kendall, I'll be delighted to meet your friends. Where are you phoning from?'

'Yass. We should be there in fifteen minutes. Don't go to any fuss. A cool breakfast will be fine.'

Wearing her soft-green slack-suit and grey sandals ready to greet her son's friends, she snipped a rosebud from the circular front garden. Standing on the verandah she turned as Harriet spoke.

'Both the front bedrooms are ready for your guests. There are fresh towels and washers on their beds. I'll bring in their small cases. Kendall can carry their heavy luggage to their rooms.'

'Thanks Harriet, he looks tired. It's not a long journey down from Sydney. We'll soon know why.'

Kendall alighted from the sports car to assist his lady friend.

'This ground is a bit uneven Brianna. Drought pocked, take care walking up the steps. I'll be with you, once I help your elderly gentleman over to them.'

An eager smile and the rose greeted her young guest.

'My son didn't tell me who to expect. Please come indoors. I am at a loss of your name, my dear.'

'Mum, this is my friend, Brianna O'Shea. And this gentleman is her priest. Father Brady, Brianna and I are exceedingly tired. After a forty-hour flight from Ireland to Sydney we're exhausted. It took ages to come through Customs at Kingsford Smith Airport at six this morning.'

'Now you're confusing me. Flown from overseas? Kendall, don't let your friends stand there in this heat. Brianna come on through to the lounge room and make yourself comfortable. Son, please show Father Brady to his room, the one opposite mine.'

Walking down the main hall, Gigi closed her office door. It was seldom locked unless she was away, working the sheep or harvesting crops. The house windows or external doors were secured every night. Her office was out of bounds to Ian at all times. She trusted strangers. Only if they proved to be trustworthy.

Curiosity stretched her patience during breakfast.

'Kendall what are you hiding from me? Why did you travel overseas to meet Brianna and Father Brady? I thought you were working in Sydney Hospital.'

'We enjoyed breakfast, so don't be inquisitive, or you'll spoil our day.' Under the dining table Kendall held Brianna's hand. 'Mother dear, we,' he hedged deliberately, 'we might have another slice of cheese. Pass the board, please.'

She handed Brianna the cheese knife and gasping with astonishment she looked at her finger, *I should reprimand him for being deceitful.* 'Brianna your engagement ring is beautiful.' Flexing her finger, a glimmer of light reflected a myriad of colours in Gigi's violet eyes. 'This ring signifies a lifetime of happiness with my son. I'm thrilled at his prospect of settling down with the woman he loves. We are honoured to have you both as our guests. We should celebrate your engagement with a party on the first of November.'

Kendall looked at his mother suspiciously.

'What made you say that day, Mum? The first of November is Brianna's birthday. I didn't mention it on the phone.'

'A woman's intuition, son.'

'My mother's a whiz. Come with me Father George, and I'll carry your heavy case and overcoat to your room.' Kendall walked with the elderly priest up the main hall and hesitated when a snarling growl came from the small hallway.

'Who asked you to come home? I didn't. And who are those people, Kendall?'

Fuming he turned and faced his father.

'These people are my guests. Mum has invited them to stay here. If you object, see me in my room. Otherwise, refrain from passing your rude comments.'

Floundering for words, Gigi glared at her husband.

'Ian, go and finish dressing. It's undignified to walk around the house in torn shorts and that grubby singlet. These people are guests in our home. They have travelled overnight from Ireland. Please respect their dignity and do try to be courteous.'

Humiliated by a woman giving him orders, Ian returned to his den. Mumbling something incoherent, a chair went flying and crumpled bedclothes collided with the door.

Relieved, Kendall then assisted his fiancé's priest to his room to administer his medication.

'Hours of immobility on the plane and in my car have depleted your energy. You need to rest until lunchtime. Your pulse is rapid. Put this tablet under your tongue Father and let it dissolve. This drug will gradually ease the tightness in your throat and the pain down your left arm.' He went to close the blinds and Father Brady objected.

'Leave them, me boy. God gave me eyes to appreciate the beauty of this great country. I dislike the room darkened. Confined in me tiny cell fa weeks in Germany, I almost suffocated. The narrow window-grill above me bunk only let a slither of air in for me ta breathe.'

Closing the door gently, Kendall walked back to the kitchen where Harriett and her daughter Rose were eating their hot breakfasts.

'The fruit from our local orchards looks yummy.' He helped himself to a handful of cherries and a greengage plum.

'Kendall, the two bowls on that table are for your guests. Would you mind carrying them to their rooms, please? There's more plums, peaches and apricots in a glass bowl on the lounge room table. You and your guests are free to help yourselves. Rose and I are going to Yass shortly. If your mother needs anything at the chemist, or wants us to collect her dry-cleaning from Biracks, ask her to let us know soon.'

'I have a pile of shirts and Brianna has several frocks. I'll take in the fruit and grab our soiled clothes while I'm there. Father Brady's clerical suit needs steam-pressing.'

'Congratulations Kendall, your choice of choosing Brianna as your wife is the best news I've heard in ages. She seems a very pleasant girl. What does she do for a living, back home in Ireland? She began to tell me, before his nibs bellowed from his hallway.'

'Typical of the prig, Harriet. He delights in humiliating me in front of strangers. He disliked mum challenging him about walking around half naked. What really antagonised him was how she told him to be civil in front of my friends.'

His mother stood undetected in doorway. 'What did you say about me being civil?'

'Mum, you shouldn't be an eavesdropper. You only heard part of my conversation with Harriet. I don't need to be chastised like a kid. Brianna will be twenty-one in December and I'll be twenty-seven this weekend. There's only one snivelling grouch in this family. And you know who I mean.'

Brianna looked from Kendall to his mother.

'Excuse me, Mrs Ross. This hot kitchen needs a revolving fan, like the one in my room.'

'Brianna's right Mum. What have Harriet and I been telling you for years? Tomorrow choose a fan from this brochure, and before nightfall it will be installed, along with new ones in every room of this house. At my expense. No arguments, Mother dear.'

'Why would I argue, Kendall? Suffering in this unbearable heat for weeks, only a fool would object to new ceiling fans. I can afford to pay for them, son.'

'I said this is my shout, Mother. The fans are my way of saying thanks. If you, Doc Jarvis and Bjorn hadn't supported me financially for nine years, where could hang my professional shingle? My official certificate states I am a qualified physician. Once we've changed, Brianna and I are going for ride. I'll be astride Potchkin. Can she borrow your horse, or one from the stables? We bought new saddles in Ennis, her hometown.'

'Yes son, she can ride Wallaby Down anytime, if she asks me first. Father George also comes from County Clare in Ireland, so he told me on his arrival.'

10am Next day: The first two hours were filled with sorting linen, cleaning and dusting bedrooms for more guests. Rose and her cousin Jasmine were exhausted by lunchtime. There never seemed enough time to do all the household chores. The new fans proved a huge relief on sweltering days to the women working in the kitchen. Without Jalna's staff, Gigi would've been at her wits end to get everything done. Ian was less than useless and she suspected this was purposely orchestrated. A constant state of inebriation kept him under the blankets and out of everyone's way. This allowed things to run smoothly as each hour drew closer to her son's birthday and the engagement party.

Gigi couldn't cope with Ian interfering. She was elated when he refused to leave the den. In all probability he would revel in breaking her fine china as a deliberate ploy for refusing to sleep with him. She vowed never to again, while he imbibed.

4 pm. With five minutes to spare, Gigi sat on her bed to admire a photo taken on the day they had purchased Abergeldie. Her brilliant floral arrangements and sewing talent had allowed the homestead to develop its own character. In the past she lacked the chance to develop her skill in decorating. Immensely proud to call this home her own, after decades of struggling to balance Abergeldie's financial accounts, a surplus of money induced her to concentrate on vital issues. It lessened her stress, of Ian's inability to cope with problems.

Her son's future with Brianna was her main concern. *I'm proud of Kendall. I think Brianna is a sweet young lady of whom I've grown so fond.* A gentle tap on the door distracted Gigi. 'Come in darling. I was just thinking of you. Sit here on the bed beside me, Brianna. You look a little flushed.'

'Kendall is unsaddling your horse. Wallaby Downs is a magnificent animal, and he's well groomed. He never shied when a snake crossed his path. I was a bit scared, until he whinnied and the snake slithered under a rock. Kendall said it wasn't a poisonous viper.'

'A physician, one of his tutorials covered reptiles and their poisons at a university. I'm going be blunt saying this. Have you studied dancing, or are you employed in that field at home? Why I asked is, look at my dressing table mirror.'

Brianna gasped within delight. 'Oh, your ballet shoes are beautiful. May I call you Gigi please? Missus is so formal.'

'Yes, of course you can Brianna. You're a sweetie for asking. In my youthful years I was a ballet dancer and sang La Scala and other operettas in Milano. During the last war my manager, who is still one of my treasured friends, and I entertained French and Allied troops on the battle fields. Accolades keep ringing in my ears on still summer nights, I relive those precious moments when I listen to Pierre's tapes. Wonderful memories are recreated by scientists in this brilliant era of technology.'

A sentimentalist at heart and with tears in her eyes, the pink ribbons of Gigi's point-shoes, worn flat at the toes, disturbed Brianna. Embarrassed by weeping, she couldn't think how to respond.

'Have my silly meanderings upset you, my dear?'

'Not really. You took me on a long, sad journey in a few words. I lost my cousin, who was an understudy to the Sugarplum Fairy in a production of The Nutcracker. Elicia died of an aneurysm. None of her family, or mine suspected that she was suffering from it or a heart defect. I seldom think of those tragic days, unless something occurs to stir the memory. It happened a year ago in Paris.'

'How can I express my sorrow over your loss, Brianna? What a traumatic period in your lives. I can't perceive how you managed to survive under severe pressure and stress, and it appears you suffered greatly at the time.'

'Not long after her death, I visited the Whitsundays and met your son. Kendall and I got on famously. He proposed to me on a yacht, as white-tipped soft-green waves broke on the sandy beach as a vibrant, rose-tinted glow appeared around sun at twilight.'

'Now I feel cheated. Kendall didn't say a thing to me about holidaying on Hamilton or Daydreams Islands. What ran through his mind is anyone's guess. You were telling me how your style of dancing developed, until we travelled on a road loaded with sadness.'

'My repertoire could never equal yours, Gigi. Irish dancing was derived from Celtic history. My shoes are either soft-soled or metal-capped.

We dance with both arms straight and our palms tucked tight to the outer thighs while pounding the floor to a rhymical beat. In full swing, everything within cooee shudders.'

Gigi couldn't help grinning.

'My son been your tutor? Cooee is a dinky-dye Aussie saying. Brianna, you've mastered the Aussie twang and it sounded enchanting. A lot of overseas travellers mispronounce the word Australia. I did, until a friend taught me the fundamentals of speaking correctly to connoisseurs at an equestrian dinner in Melbourne.'

23

6 am. The day prior to the engagement party, Bjorn and his crew began erecting a huge marque on Jalna. This eliminated tempers, fierce or otherwise, from accelerating to flashpoint because of Ian's irrational moods. He and Kendall were constantly at loggerheads, nitpicking over insignificant and unimportant issues.

Several interstate shearers had offered to be relief workers. After giving Jalna's top ringer a demonstration of their skills, Lanky suggested to his boss that they should be put on Abergeldie's payroll.

Within a day or two their new foreman was due to move on site with his family. His wife would alleviate the intense pressure off Harriet Graham, the homestead's present cook-cum-housekeeper.

This huge property comprised four bunkhouses for full-time residents and household staff. All seasonal workers and stockmen were housed further towards the shearing and woolsheds. Yet still within cooee of the homestead.

The large rain-proof wooden structures housed the shearing equipment, plus lengthy tables on which the fleeces were thrown. Once each fleece was sorted and scoured it was gathered and bundled into white, brattice bags. Both these sheds were conjoined by a breezeway and had their own access entry which fed in from the highway. This unsealed road bypassed Abergeldie's main entrance by a hundred metres and circumvented rows of poplars. In summer the trees provided a windbreak. Lined up behind them, was a grove of conifers that protected the homestead from vicious westerly winds.

Bjorn wished his spread was strategically placed like Abergeldie. At this late stage he wasn't prepared to invest in new buildings until they needed restructuring. 'The stables and woolsheds are sound enough to protect my horses and the men from bitter draughts in winter and a scorching summer sun. Why waste my hard-earned funds to rebuild the lot now? I will when time and age necessitate their restructure.'

The day before Kendall and Brianna's party, Ian confronted his wife in her room, which was still out of bounds to him.

'Why have you refused to hold the party here on Abergeldie? Isn't my home good enough, Gigi?'

She refused to comment, yet thought, *Our home, indeed. You skinflint. You haven't contributed a cent towards the upkeep of this house, or running of its sheep for a decade.*

'With our septic tanks full, Bjorn suggested a barbeque on Jalna. You needn't worry, Ian. You won't be footing the bill. I have outlaid for the catering and so has Kirsten. Our treat to Kendall for proposing to the woman of his dreams. Even you must admit that Brianna has brought laughter back to this mausoleum.'

'You might think that. But I disagree. How can I laugh with that miserable crud objecting to everything I say? If he dropped dead, or sprouted wings and flew off the planet, nobody here, or in Yass would be any the wiser, nor would they miss him.'

'Now you're being childish, Ian. What a horrible thing to say. Kendall is your son.'

'That is debatable. I didn't sign his birth certificate. You were in Llangollen for six weeks. Who did you invite to your bed? I've often wondered if *he* is my son. He's an ignorant prick, who doesn't care a damn about you. Wake up from that fantasy world you've conjured. The first chance he gets, he'll drop this Irish bitch and seek a woman of his miserable equal.'

'Why are you so bitter, Ian? Your mind equals a sewer rat's intelligence. They have more decency in their whole body than you have in your brain. You disgust me, Ian.'

Before sun-up the following morning Ian rode down to the woolsheds. Abergeldie looked magnificent in the fragile light of day. This heritage-listed building far outshone Jalna's weathered timbers and even put the distant magenta hills to shame.

7 am. Two freshly iced cakes were ready for Harriet and her daughter to ferry over to their neighbour's kitchen in her son's car. Its boot

contained gifts for the engaged couple, plus a complete change of clothes for Kendall.

Bjorn directed Harriet's son Paul to park in close to the breezeway.

'You can leave the car there all day. My men have finished erecting this stupid contraption. Give me a hand to whack in these last two tent pegs, Kendall. I keep hitting bedrock, or something hard with the sledgehammer and this damn patch of ground won't yield.'

'Put a bit of muscle behind each strike, weaky and the peg should go in okay.'

'A weaky am I, mate? I'll show you who the weakest bod is here.' A hefty swing and the hammer miscued. Bjorn swore as the iron sledgehammer struck the toe of his left boot. Jumping on his good foot he bellowed in pain. 'Struth, that hurt. I think me big toenail's split asunder.'

'Sit on this wooden keg. Once your breathing settles, I'll unlace and remove the boot. It will hurt, so grit your teeth BJ.' Kendall eased the boot free. 'Wiggle each toe in turn. Good. No broken bones. You'll survive, but only if you stay off that damn foot and keep the leg elevated.'

'Shit, this would happen today. There's a ton of work to do before six. What time are your guests due to arrive over home? Bet your old fart won't turn up tonight. Ian always makes himself a bloody nuisance. Even when he's sober.'

Kendall passed a wooden stake to his neighbour.

'Lean on this BJ and hobble over to that wooden seat. I have a firm hold on your left arm. There's no chance of you falling. Take one step at a time. If you try to gallop, you will collapse. Don't hurry.' Kendall's free arm reached for the iron-laced bracket, supporting a basket of multi-coloured petunias. He toppled, but his fingers clung onto the verandah's wooden post.

'Struth, that was too bloody close for comfort. Mate, if you had taken a dive, you'd have pulled me with ya inta those hot coals in the barbeque pit.'

'Quit giving me a mouthful of cheek, and take it slowly up the steps, Bjorn.'

Aided by the young doctor, BJ hobbled indoors.

Kirsten slid a kitchen chair under her husband's bottom. 'What have you done now, Bjorn. It seems I can't trust you out of my sight. Stay there and don't make excuses.'

'If you must know, the f … err … sledgy slipped and struck me hoof. I'm not dead, or severely injured. Fetch me a beer and clamp it, Kirsten.

The doc here has ordered me to keep this foot well above me heartline. What good that'll do, I'm buggered if I know.'

Bjorn Svensson dreaded the idea of Ian Ross coming tonight. The antagonism between himself, Kendall and his father would inflame a delicate situation and could create another catastrophe. Abergeldie's demonic ex-owner's declaration of war against his neighbour increased daily. Only the wind heard Ian's cry of revenge.

10 am. In a sombre mood, he leant against the front verandah railing to survey Jalna from a distance. Gazing across its home paddock at Bjorn's men loading the last of their tools, his mind contrived a quagmire of melanistic meanderings. *After the party tonight that argumentative and self-opinionated bastard will have nothing more to do with my family. Nor will he interfere with running my mob of sheep or the staff here on Abergeldie.*

Touching a flame to his cigarillo, Ian quit thinking and snarled.

'From today on, Gigi will take orders from me only. If the bitch condescends to sleep in our marital bed, I won't raise a hand to her. But if she ignores me again, I will beat her arse until it bleeds. To segregate me was a cruel move on her part. If I desire to be assertive with my rights as a husband she daren't rebel or refuse my love. And should she try to crush my desires, I'll have my revenge by forcing her to submit to me under a cold shower.' In a devious and manipulative mood, this evil ex-Nazi didn't consider rape constituted debauchery and a violation of his wife's right to privacy.

Before breakfast the previous day, Bjorn had decided to leave a skeleton staff on Jalna. Accompanied by Bernie and with Lanky they began to get everything underway. With the season over, most of drifters had stayed on to assist in organising Abergeldie's relief shearers to become acquainted with their new jobs.

After a test run shearing ten sheep in the home paddock, each man had proved his worth. Lanky Hancock reported to Abergeldie's relieving foreman, Bruce Webster.

'All the permanent shearers are proficient and their standard of work is okay. It's good to see they've left this shearing shed and them wool tables in an immaculate condition, Bruce. There's still a hint of an oily aroma emitting from the wide timbers.'

'Yeah, the cracks between those floorboards are loaded with lanolin, a natural oil which exudes from deep beneath the sheep skin. Thanks for overseeing my men, Lanky.'

He silently appraised how clean the shearing shed looked.

'I intend to see they leave both sheds in a pristine condition. The smell of lanolin brings to mind my years working as a shearer down in South Australia. They were the days of carefree enjoyment with my wife and family.' Bruce then thought, *I'm not accustomed to toadying to the habits of a drunken bum. From what Arthur Hancock just implied, it seems Ian is addicted to drink.*

'Hey mate, your shearing sheds outshine mine on Jalna. This building's been restored to its former glory. Ya can tell that, by the fresh stain on those beams. The floorboards are solid by the way they sounded when I trod them down. The wool sorting tables are less worn, compared to mine. Lanky, my top-gun, is a bonza bloke and a hard taskmaster when supervising our younglings over home. If frustrated or angry, he can swear worse than any soldier did on the frontlines in Europe during the last war.'

Arthur Hancock flexed both eyes towards the structure he called home. Jalna was the only home this weather-beaten shearer, approaching fifty, had ever known. Not unlike its boss, whom he considered more of a brother than an employer. They seldom argued and never a cross word passed between them, unless Ian Ross caused a row over something trivial.

Earlier in the day. Bjorn couldn't afford to pull his blokes from sheep crutching, or their regular jobs to do the menial task of fixing tent ropes to its metal stays. Not when he and Kendall had coped well, until his barney with the hammer. Retying the saturated bandana around his sweaty neck he sighed. 'Bet you're looking forward to tonight's bash?' They both looked at the stable marquee erected on Jalna's back lawn which would stay unless the predictable wind sprang to life.

'Yes, I am rather. And so is Brianna. Some of my friends from uni arrived last night and they're staying in a Yass hotel. Apart from them, it's a family affair.'

'Thanks!' BJ thought this was too good a challenge to miss.

'For twenty-odd years Kirsty and I have considered you our family. Kendall, do you mean that we're not invited to the barbeque on our own property? Come on mate, be honest and admit the truth why you're snaky with your old fart? I'm a patient listener, most of the time. Unless I'm riled by some idiot.' This solemn reply came with a wry smile.

Bjorn spluttered as a willy-willy spewing dust encompassed them. He copped the lot and wondered why Kendall, who anticipated it, had turned away. Cursing, he spat out a mouthful of grit and foraged in his overall pocket for a clean sweat rag.

'Pass me one those middies, Kendall. I'm dying from lack of an ice-cold beer.'

'No Bjorn. Beer contains arsenic, an inhibitor of growth. It will make you suffer the agonies of hell. Keep off that damn foot and let the swelling subside. Drink a glass of fresh orange juice. Helen's worked her butt stiff, gathering fresh fruit from our orchard. Apart from an hour of squeezing four jugs for tonight's bash.' The angry scowl he received from his ungrateful patient, made Kendall furious. 'You'll only have yourself to blame, if you force Jarvis or me to remove that blackened toenail. And you know I'm right, BJ.'

'Shit! Why must you be so damn assertive all the time, Kendall. It's time you went home. I need a cool shower, to be rid of this damn sweat.'

'Don't you wet that foot. Wrap this waterproof bandage around it. Park your arse on the bath, while I wash everything, bar your genitals. You're big and ugly enough to do that. In my profession, I've injected loads of naked bums. Get in that bathroom now, Bjorn. And I'll give your back a good rub down. Then you can towel yourself dry.'

'You listen to me. I *am* not a horse. I do not need a rub down or a roll in sand. You can bathe me balls and the parts I can't reach, Doc. Let's get moving, before me brain goes wonky.' This self-appointed patient tried to sprint to the bath and tripped over a stool.

Foreseeing this mishap, Kendall couldn't prevent it from occurring. Precariously balanced, Bjorn slipped on a bar of soap and fell headfirst in the bath. Then drying his private bits, he let fly with a string of unmentionable swear words.

'Come on, pass me the towel and stand still, until I've dried your shoulders down to your bum. Hold the towel-rail, then you won't topple over again.'

'Because you're a qualified quack, don't give me orders, mate.'

'If you didn't appreciate my joke, I'll go home. Whatever you do, cut out this mate nonsense. It's slang and reflects on my good character,' Kendall declared half in earnest, yet bordering on jest. This word with its sarcastic inference, now taken out of context, he considered unworthy of his professionalism. His father, who derived great pleasure in belittling both he and Bjorn, often mimicked this saying when in a narcissistic mood. Kendall detested the word mate, and considered himself above resorting to use it or any other Aussie slang.

'Seriously, neither Mum nor I would have survived without you, Bjorn. You and Kirsten are closer than *his* family to us. Mum has never seen any photos of *his* parents. Brianna has noticed the friendly and genuine rapport between us all.'

'You've struck gold with that lassie. Brianna's a charming young lady and she idolises you. She seems quite fond of your mum. It's early days so don't bank your hopes on having kids for a while. I reckon she's a pretty good looker. Some Irish women show their rage long before they reach thirty.'

'What a weird description of my fiancé!' Kendall laughed. 'She is a lovable girl, and a real treasure. Mum welcomed her with open arms.' This frivolous mood brightened a dismal day and he grinned. 'Bjorn, I cherish Brianna Skye and I will miss her.'

'You'll miss Brianna, why? Maybe my joke was untimely. You're an ideal couple. Is she flying back to Ireland this soon? I assumed she would be here for a month, or longer. Kendall, is it wedding bells I hear, or haven't you set the date?'

'You'll all be told at the appropriate time. Brianna wants to be married in a private ceremony on Abergeldie. You can imagine what that old bastard would say, Bjorn. I wouldn't put it past him to try to seduce Bridy. Mum overheard him on the phone saying that Brianna was beautiful and she'd be an easy target to bed.'

'Has he spoken to or seen her close, Kendall? Your mother gave me the impression that neither Brianna, nor her priest have been approached by Ian.'

Closing his eyes, Kendall shook his head. Even the thought of his father attending the party made feel him ill. And he knew the consequences if they married there. 'The maniac will never consent to our wedding being held on Abergeldie, which he thinks still belongs to

218

him. I hope the crank falls and breaks his neck before this evening. See you around seven. Mum asked me to drive my car over. It will give her chance to relax and have a few drinks without the fear of being locked up on DUI charge.'

5.30 pm on Abergeldie. A twist of both shoulders revealed a deep cleavage in her bedroom mirror. *This frock blouses in all the wrong parts. Brianna is fashion conscious. I'll ask her opinion on this rag.* 'Darling, are you decent? May I intrude on your privacy?'

'Please come in, Gigi. You look a little flustered. Oh, the frock? It doesn't suit your complexion. If I take a peep in your closet there may be something to compliment your violet eyes. A gentle mauve or a delicate green I think for a casual soirée.'

'My lilac blouse goes with this grey crepe skirt. You can pick which one would be best to wear on a hot summer's night. My silver or grey flatties are comfortable. Not high heels. Be my luck to trip or fall over something.'

A quick rummage through a wardrobe bulging with clothes revealed a long frock of lavender. 'This gown with a silver thread woven though the fabric, and a twelve-inch split in the back is ideal. Try it on Gigi and swirl towards me.'

'It still fits me and the draped neckline feels soft against my skin. Damask is a cool fabric on hot nights, it breathes.'

'Wear it tonight with pride. Kendall asked me what I have chosen. I told him not be inquisitive. I packed only two cool dresses. A sage linen and my pastel-green cotton has puffy chrysanthemums brocaded in white around the hemline. You can select which I should wear.'

The pale green reflected in her sensual, soft-grey eyes. A spiral knot of auburn hair crowned her unadorned head. On her pillow, Brianna found an evening clutch. 'This bag looks new.' The mystery deepened. Unclasping the metal knobs, she discovered a return slip. 'Kendall bought this for me. He knows I left my change purse at Rianna Airport.' A shrug of her shoulders as she stepped in the bathroom left Brianna guessing as she donned her shower cap.

24

After being away for an extended period and depleted of funds, due to falsifying the amount in all his cheque books, Ian had no recollection of signing Abergeldie and its landholdings over to his wife. A philanderer, he'd grown restless in Melbourne. Tired of sowing his sexual seed in the fields of femininity in that busy city, he knew it would only be a matter of time before moving on again. Wanderlust had devoured his energy to cope on meager funds. Nobody there cared, nor would they miss him.

As Christmas drew closer Ian found cheap grog his only comforter. Chasing the dragon was causing this man who tinkered on the brink of a grave. Inability to think constructively with a dysfunctional mind, he imagined everyone was plotting his downfall. Inebriated most evenings, there were lucid periods during the day when his mind reverted to sane and logical thinking.

'Blast the lot of them. Svensson's a smug bastard and thinks he knows everything about my life. Even the thought of going tonight makes me want to puke.' Sober, Ian conducted all overseas business deals from his den. This area was forbidden territory to his wife. Neither she, nor time could prize him from his self-induced exile and comfort zone. The passion for writing his memoirs increased with each drawn breath and grew to become an obsession.

A drink-ravaged brain failed to recognise his wife as a contention of annoyance. He considered Gigi a mere sex object and toyed with her emotions. Though of late, his motivation for sexual interludes had somewhat waned.

This she readily accepted. To challenge why he had spurned her, no longer worried this fragile woman with a heart overburdened with anguish. Having obtained her sexual freedom, she certainly wasn't going to endorse her signature on a divorce document. She sensed this idea was a temporary mode of uncivility of Ian's.

A snake doesn't shed its old habits when shedding its skin, she mused. And with her mind on a higher plane, she relaxed to sip a whiskey-sour in the lounge room, until her son and his fiancée were ready to leave for Jalna.

Sweating for hours under a hot sun while erecting the tent on Jalna had zapped some of his strength. Resting his bottom on the priest's bed, Kendall recalled how Bjorn had mopped his brow. He smiled and flexed his shoulders while meditating. *That work today was backbreaking to the uninitiated. How we coped okay in the heat I'm damned if I know. We were fools to work in those extreme conditions. If Bjorn doesn't keep off his feet tonight, the damaged subcutaneous flesh under the toenail will bleed and he'll lose it.*

'Oh dash, me missal's fallen on the floor …'

'No, I caught it Father. There's something tucked in the spine of this prayer book.'

'Before we retire tonight, I'll be tellin ya a sorry tale why I keep it there…'

A feminine voice interrupted their conversation and smiled knowing why her priest always dropped the letter 'G' when speaking. 'Kendall, your mother is ready to leave and so am I. Please walk Father George to the car. Or we will be late. Your father isn't feeling very well tonight and I doubt if he'll be joining us on Jalna.'

'Where is he now, Brianna? He roared at mum a while back. She wanted to take his photo and he threatened to smash *my* camera. Knowing his objection to the heat, he'll be under a cool shower. Keep an eye on my medical kit and stay with Father George, please darling. If Mum asks, tell her I'll be back soon.'

Kendall was furious. Snatching his suitcoat off the bedpost he hurried to see what his father's excuse was this time. 'Probably a ruse to delay mum, or he doesn't want to attend the barbeque on Jalna.' He didn't knock, just barged into his father's bathroom.

6 pm on Jalna. Bjorn gestured for Cory and his brother to stop turning the barbeque rod. 'You boys have done a good job roasting the pork. Two

of my men will take over now, to give you boys a spell. Wash your hands and get some tucker. You've both earned a rest.' He then spoke to the young muckers manning the lamb-spit. And repeated the order.

Reminiscing on the back verandah seat a cool breeze revived a work-fazed spirit. Bjorn looked at the setting sun and sighed. *I meant to tell Kendall he's kept our hopes alive for a son. Our own child, if we had one, could never match his generous carefree nature and sincere personality. My family have worked hard to maintain our friendship down the years.*

Earlier both their minds had tripped along the same parallel and Bjorn now recalled how his young professional friend could only have benefited from studying medicine, because of his effort to succeed in life.

'Kendall is aware that neither he nor his mother would be living on Abergeldie, if our friends hadn't rallied to prevent the bailiffs from confiscating everything and declaring their family bankrupt. I better go and shower before Bucknell comes in with his secretary. Peter Bucknell, a semi-retired diplomat, is a dedicated parliamentarian, Kendall and his mum will meet him tonight. Boy oh boy, is he a handy bloke to have around when a man's in strife. He's a genuine bloke, straight as a dye and he doesn't mince words.'

7 pm. About to disrobe in their bathroom, Bjorn pulled a roll of peppermints from his dungaree pocket, blew the chaff off and handed a mint to Kirsten. 'Don't just stand there darls. I need some privacy. Oh Kirsten, is my blue casual shirt dry? I threw it and my navy dacks on the washbasin. Do I need to wear a tie tonight? I hate neck-chokers.'

'With an *open-necked* shirt? No of course not. You're not the MC tonight. You must keep off that swollen foot and do what the doctor ordered. And no maybe's or but's either, Bjorn.' She glared at the clean floor. 'What's that chaff doing there? Why can't you be more careful? Use your wet towel to flick it under the console. Helen's busy and I going to finish dressing. You really are hopeless.'

7.30 pm on Abergeldie. Ready to tackle his father and munching on a peppermint Kendall slammed Ian's bathroom door.

'What do you want? Come to snigger or growl about me not being ready?'

'I don't give a damn if you're coming or not. I will not tolerate you speaking to my mother, or threaten her like you did a while ago. You really are an inconsiderate bastard and you can go to hell for all I care.'

'I'll go over when I'm good and ready,' Ian retorted sarcastically. 'I dislike that priest. I recall seeing him somewhere before, yet I can't think where. Shush, that's him talking to her now. What the hell are they doing in my den?'

'You are disgusting, standing in there naked. Wrap this damn towel around your lower half.' Kendall threw it. 'Cover your genitals.'

'You barged in here. Now leave or you'll cop my fist in your face.'

'If you thrash mum ever again, I'll kill you.' Kendall ushered his mother into the hall. 'Cover your genitalia with the damn towel. Mum was right. You are a despicable wretch. Thank God you won't be upsetting her tonight. It will be peaceful without you sticking your damn nose in everyone's privacy. Now get dressed.'

'Get out of my bathroom and my house. Who do you think you are, giving me orders? You always were a snivelling brat.'

Kendall threw the bar of soap at his father.

'Your house! You relinquished that right by forcing my mother to sign everything over to you. In case you've forgotten, let me refresh your memory. Three months ago, after you abused her when you were legless, Bjorn came over here and forced you to sign a deed of title. That letter is in my safe deposit box at the bank. Only mum or I can sign for its release. You own nothing in this house. Not the sheep. Not the tractors or the woolshed and bunkhouses. They belong to my mother. The clothes in your wardrobe, the things in the den are all you can claim. Oh! And your horse. What a pathetic creature she is. A hack ready for the knackery.'

Disregarding the previous warning Ian's raised fist collided with his son's shoulder. Kendall ducked and his arm had shielded his face. A return, swift uppercut collected Ian's nose. 'There's another thump waiting, if you step out of your den before we return from Jalna in approximately five hours. I'll give you five minutes to get in there. Remember, I have the duplicate key of your study.' The bathroom door slammed and then rebounded.

'Move back please Mum and let me pass.'

'Kendall, your shirt is splattered with blood. Did you strike your father?'

'Yes, I did, he deserved it. I'm going to change my shirt and trousers. Do not go near him or the den. He's psychotic, a raving lunatic. The man is certifiable.'

In an effort to calm Kendall's mother and the situation, Father Brady touched her hand. 'Harsh confrontations cause dissention in families. I try to avoid arguments. In the last war a horrible Nazi captain ordered his men to whip me. Twice he questioned me and ordered his thugs to break me spirit. I survived because a priest saved me. Father Patrick put his clothes on me. Those beastly Nazis then shot Patrick, instead of me.'

'How terrible it must've been for you. Do you recall the Nazi's name, Father? His face must haunt you daily.'

'He does, in me dreams. His name, now let me think? Captain von Breusch was the cruelest officer in the Berlin Chancellery. I shan't ever forget, nor forgive him.'

'Come, walk with me. My son and Brianna are ready to leave. Try to dismiss all the horrible memories from your mind and enjoy tonight. Oh, I forgot to wrap their presents.' Gigi then remembered the presents were already in the boot of his car.

'Careful Mum, these front steps are slippery. I'm not grumbling. We need the rain, and it was a good summer shower. The roses look fresh and vibrant tonight. Not wilted and dying of thirst, like me.'

In the den, Ian stuffed his briar-pipe with weed and tramped it down with a pencil. 'No, it's impossible. That Irish voice cannot belong to the priest whose death warrant I signed in forty-two. Or was it forty-four? From my office window I instructed the firing squad to aim, fire and kill the Irishman. I only caught a glimpse of him in the main hall on his arrival. I now know there *was* a traitor amid my subordinates at the Chancellery. The priest's declaration means he must have some proof that he hasn't disclosed to that bastard son of mine. Where has he hidden it? I can't afford to waste time searching either of their rooms. Pity my plan to seduce his red-headed fiancée tonight while everyone's asleep must be aborted. I may not get another chance.'

Overwhelmed by the priest's admittance of exchanging clothes in his cell, and his miraculous escape to freedom floored this Nazi absconder. *I would never have believed his disclosure of that event if I wasn't standing behind this door. My ploy to seduce the miserable crud's fiancée must be abandoned this evening. I need to concentrate on writing a plausible excuse, one which those idiots will accept as truthful. Pen and ink on paper don't lie. Why bother signing the note? No, initials will suffice.*

Gigi joined in the jocular spirit and enjoyed chatting to Brianna and her priest in the car. Acknowledging his graceful smile, she reciprocated with a bigger one.

'Father, I hope the flight from Ireland wasn't stressful? I did it a few times in my younger days. Long hours of immobility can be exhausting.'

'I'm pleased ta have met you, Mrs Ross. Me darlin here mentioned ya in her last letter.' With pride, he waved it aloft. 'Such a nice boy she'd be marryin, Kendall could only have a charmin mother.'

'You're a flatterer, Father,' Gigi responded, her face blushing a rich puce. 'Brianna told me that you are a friend of her family in County Clare. I lack knowledge of Ireland's topography, so I'm not conversant of the counties.' She apologised to the priest, 'Father, ask my son to wind up his side window. There's a cold draught in this car. It could be coming through those circulation vents.'

'The gifts I left on the lounge dear lady, they're from Brianna Skye's family. The small one I made fa Kendall's birthday. Her folks back home will phone sometime tommah.' The armful of brightly-wrapped gifts he'd given to Brianna sat on her bed with Kendall's engraved wristwatch.

The priest's Celtic brogue fascinated Gigi and in the fresh, country air it sounded more pronounced than ever.

Love being gentle and blind, it instilled in Brianna the best way to settle their tiff while standing in Jalna's breezeway. 'Darling, look on the positive side of your argument and smile. His rudeness wasn't necessary or worth fighting over.'

'No, you listen to me Bridy. I won't be a hypocrite and acknowledge who he is to anyone. To me he doesn't exist. One day I'll tell you why I've turned my back on him. Then perhaps you'll understand why I am so bitter. I will not allow him to come between us, or spoil our evening. Leave it at that, please honey.'

'Don't be angry with me Ken. I love you too deeply darling, to keep arguing with you. It's futile and *we will* spoil everyone's night, without him ruining it.' A brief moment was stolen for a kiss. 'We haven't long together before we go our separate ways, and heaven knows for how long.'

Cuddling her, Kendall asked quite calmly.

'Normally I don't believe in love. Often, it's a fallacy of lustful desires that some men think is their God-given right to pursue. Since meeting you, I've become more malleable in my opinion. I sincerely care for you Brianna. What time does your flight out for the UK leave from Sydney?'

'Eight, your time on the Monday after next,' she replied with tears lingering in her eyes. 'It means I'll have to leave here on the Sunday. Early morning flights I detest. Still, it can't be helped. I'm booked into an exclusive hotel not far from the airport, or so my travel agent said. I'll check the time when I ring Sydney to verify my flight. I only have a week then before my contract begins with the Irish Dance Company. Our time together has been all too short this break. After years of tedious studying you need a holiday.'

Kendall frowned. *I was sure Bridy was flying out after me. Oh, I must've got the time wrong or misunderstood her flight schedule. I must verify my overseas flight today.*

After a moment's hesitation Brianna enquired, 'What time does your flight leave Canberra for Sydney, Kendall?'

'Looks as if I'll be pulling out before you, on that Sunday morning. We can go across to Canberra with my mother. Then she can drive my Jaguar home. I don't fancy leaving an expensive vehicle parked at the airport for an extended period. Darls, you know how I'll dread leaving you for six weeks or longer.'

'Kendall, none of this was your fault. Unpleasant incidences happen in the best of families. Darling, we should appreciate these few days we have left. I couldn't leave home until my contract was validated for different seasons. They have granted me extended leave. My mother cried and wanted me to stay in Ennis. An impossible hurdle, one I couldn't avert. Let's enjoy tonight together. I promise not to forget your birthday next month. Darling, promise to ring me on Abergeldie every other day from Austria?'

An anguished look impinged upon Brianna's delicate features. The flood had broken its banks and tears cascaded down her cheeks. Their argument over his father had knuckled the wood. Brianna regretted it, disillusioned over her parent's inability to be here on this, their most memorable night. Sadly, it wasn't to be.

'Just being with you, is all I crave. I've wanted to meet your mother for ages. She's such a kind and lovable person. You should be very proud of her, Kendall.'

Snuggled in his arms Brianna's tears embellished his neck. As she looked up he smiled down in those misty eyes. 'Before I leave, I promise we'll have dinner in a small quiet restaurant in Yass. I don't want a fuss made on my birthday, darling.'

Brianna nodded in agreement of being able to celebrate his special day without strangers delving into their privacy. Kendall had promised to pay for her parents to fly first class from Ennis to Australia a week before their wedding. He also offered to contribute to their overnight stay in a hotel in Sydney.

His lips embraced her dewy eyelashes. 'Darling, I'll be anxious after completing your last, long stint in Ireland. The journey here has taken a strain on your nerves. Bridy, I'm sorry we can't spend more time together before we leave.'

Brianna could feel more tears building behind her eyes. Fighting to hold them back, she couldn't convey to this caring young physician how much she would miss him.

To console her, he clasped both her hands. 'Brianna, long weeks overseas will be intolerable for me. Darling, even if we're parted for an hour, it will seem like an eternity.'

Words were inadequate to express her love. 'Ken, I know with your principles we must keep the faith. You mean everything to me. Why can't we consummate our love tonight? It wouldn't be a sin, because we love one another.'

'I promised you a church wedding and on my salary I'm finding it difficult to cope. Don't forget I paid a grand for these fans to be installed. Besides, being a qualified doctor isn't all glamour. Do you realise what you're letting yourself into by marrying me?' Smiling down into her cloudy eyes, Kendall bent over to kiss her lips. 'At times work affects me. Irritability and crankiness are two of my disruptive traits. You don't need or want a grouchy husband.'

'Can't you change your plans and return to Ireland with me, darling? Then if I fall pregnant my contract will be voided.'

'Brianna, I will not reconsider what I said. Respect my principles as I do your religion. To put you in the family way would be criminal. You are very precious to me and I respect your parents. I was conceived as an illegitimate child. I don't intend going into it again now. I refuse to let a child of ours go through hell, like I have. It's the blight of my life and always will be, while ever he's alive. Forget it! My love for you reaps no bounds. I couldn't and won't hurt you by making you pregnant. Let's get out of here, before I weaken and change my mind and physically love you, my temptress.'

Retouching her nails in the lounge room, Gigi impatiently waited for Ian to hurry. Leaving his bathroom with just a towel wrapped around his middle, Ian rushed past her. He wanted to get his underclothes from the den. Ian had left them there on answering the phone. A dual line, he cursed having to wait until a neighbour had finished gossiping to his business partner in Wee Jasper, thirty miles south-east of Yass.

Shocked over the towel's brevity that barely covered his private area which a march fly hovered above, she was astounded when Ian walked to his study. *Why there, instead of going to his bedroom?* Gigi scoffed. 'I hope that black-striped fly stings his genitals.' Ready to dress him down for strutting around half naked, she heard a light step exit their son's room. She assumed it might be Brianna.

'What if she or the priest had seen Ian scanty clad? This idea is too horrific for me to contemplate. His naked body isn't a pretty sight at the best of times. With guests in the house Ian should've shown them and me some respect. What is he hedging for? If he doesn't hurry we *will* be late.' Gigi expressed her frustration with a long, heartfelt sigh.

Following Ian to his room she queried in a moderate tone, 'Are you coming to the party? Please don't fob me off, like you usually do. Don't belittle me either, by making some ridiculous excuse. Yes or no will suffice, Ian.' The fierce tone expressed her anger. She refused to linger while he condescended to reply.

Gigi smirked and recalled her deceased grandmother saying: "Live with a man long enough and his indecent mannerisms will rub off on you." *Of course,* she thought, *this metaphor perfectly suits Ian and his draconian attitude to all priests.*

Ian Ross detested priests and often boasted of the fact. Why, she never knew. But then, he hated most people whom he came across or mixed with in society. She surmised he loathed her too at times.

His obnoxious tone imaged his ugly facial expression as he waved her off. 'I have accounts to total in my ledger. Get out *now*, before I do something rash.'

Hesitant to answer, Gigi pondered. *What accounts! Kendall or I pay Abergeldie's bills, not him. Ian didn't have to be nasty with me. I've done* nothing *to warrant that horrid rebuff and his aggression.*

Hearing her son and Brianna discussing something with the priest at the front door, Gigi politely excused herself. Ian's display of uncouthness lingered on her mind. Why wait to be further abused by such an ill-tempered prig.

Kendall wasn't fooled. He knew why his mother was furious.

'He wants you to stay home tonight, doesn't he? Well, it's not on and I'm not leaving here without you, Mum. Not in the savage mood dog's body is in at present. Forget him, he's not worth worrying about. We'll all be late if you're not ready to leave.'

'Respect your father, Kendall.' Her voice took on a calmer tone. 'Give me a second to powder my nose. Fetch my coat from the hallstand, please son. I can't have your guests seeing me unadorned, or shivering in this bleak, night air.'

'Father George, I'm sorry you witnessed my argument with that horrible man until my Mother intervened. I can't apologise enough.'

Combing his hair in the den Ian's brow narrowed. *Perhaps if I'd heard more of that old bastard's exploits in the Fatherland it may've been to my advantage? I couldn't afford to miss a word or be seen by him. On recall I did hear enough to whet my curiosity.*

In solemn tone Father George had responded to Kendall.

'Well me boy, I lost a dear friend at the hands of a maniac in the Chancellery. A priest, a countryman of me own faith. Patrick saved me life by takin me place before a firing squad.' Rubbing the scar below his eye, it stirred to life painful memories of the past. Reminisces Father George had never disclosed to a living soul. God had been his witness and through His wisdom, this Irishman had survived.

25

8.20 pm. Reminiscing in the den and while reorganising his tan valise, Ian looked at his deceased father's fob watch. *How could that Irish priest have heard a volley of rifle shots, from where he said he stood on the Chancellery steps. It floored me when he declared that he'd never forget my face. It serves him right, if his memory of that day haunts him. That wasn't a priestly thing for him to admit. He did say that he hated me and that I was a barbarian. I wasn't the monster he called me. That idiot has a sick mentality. He's the one who should be certified.*

Ian gasped on tripping the tumblers of his wall safe.

'A nauseous feeling gripped my stomach. To be trapped in my own home by that brainless priest is unthinkable. This was the last thing I expected tonight. If I hadn't been standing behind my bathroom door, listening to him rambling on about his incarceration in a cell, I would be in real trouble.' Flopping down on his swivel chair he remained motionless. His face had aged in minutes. Dispirited, all colour had drained from his face. His true identity was now revealed as Rolf von Breusch. As Ian glanced at his reflection in the study mirror he looked ghastly.

Confused, he cradled a throbbing head in shaking palms. A palpating heart far out-paced his mind. Betrayed by his life in Germany it affected him greatly. The clandestine truth unveiled caused Ian to quiver at the possible consequences.

He could neither think nor ingest all the priest had relayed. His own incoherent mutterings penetrated his ears only.

'I have not travelled this far to be caught like a rat ensnared in a trap of my own making. I must think clearly. My freedom is the important factor.' Ian gasped for air. Even this small claustrophobic room closed in on him with the oppressiveness of a prison cell. It began to stifle him. *Why open the safe? Now the house is quiet I can think rationally. Time has woven me into its treacherous web of intrigue. The threaded lies sown in greed have now returned to haunt me and threaten my life. I know my wife's safe digits so I might tackle it first. I'll remove all her jewellery and her important documents. Then I'll collect my forged documents from this safe.* The pile on his desk grew higher as his wristwatch hands turned clockwise.

A relieved sigh, a strange sensation tensing his muscles relaxed as did his limbs. Ian heard the hall clock strike ten. Things still needed to be done, his mind was impervious to intrusive noises. Slowly his tenseness settled in the now quiet house. Regret faded and he focused on what item to select for an unknown future. His hearing was acute again and an enlivened spirit found its home in the vortex of an ever-whirling brain. Replenished with vitality, Ian concentrated on more important issues. Hope for the youth of his homeland to be rejuvenated stimulated his eagerness to leave Australia and travel to countries he remembered in his youth made his entire body tingle with enthusiasm.

'My first wife and Gigi both belong in the past.' Ian smiled at the vision peering back from the mirror. 'Fortified with the courage of a Reich officer, I am marching towards a creative and positive future. Nobody will interfere with my plans from this night onwards. I am now *master of my own destiny.*'

The wind's cry made him receptive to its presence. *Rain, hail and damn their party to kingdom come. I hope they all drown for the misery that lot of bastards have caused me down the years in this damnable house.*

8.30 pm. A vague look pervaded the elderly priest's wizened features as they entered Jalna's marque. 'I thought ya son and Brianna came in this tent well before us, Mrs Ross.'

'So did I, Father. Here's Kirsten, she'll ask one of the waiters to pour you a whiskey. I'll go and see what they're doing. It's rude of my son not to be here to welcome their guests.'

Father Brady looked at the elaborate decorations then his mind travelled back to the time of his incarceration in the Chancellery and horrific memories immerged. *A lieutenant I could understand takin orders. That fiend only gave orders. He was a demonstrative man, hell-bent on*

reprisals and revenge. Anyone who antagonised him walked the path of death.
A lone finger touched the scar on his left cheek. Aware of being in select
company he removed his hand and clasped both of them in silent prayer.
As they lowered one finger caught on the wooden crucifix, he'd worn on
his arrival at Abergeldie when he blessed the engaged couple.

He remembered two precious items, this cross and the priestly cassock
his lifelong friend, Father Patrick Kelly had worn on visiting his cell, prior
to his own escape from the Chancellery. On leaving Berlin in forty-five, he
recalled Father Ignatz blessing both these items before giving them to him.
Later that same year, in the solitude of his home, this devout Catholic,
George Brady was accepted as a priest in the local seminary.

9 pm. After a quick visit to the workmen's toilet block on Jalna,
Doctor Ross walked back to the marquee and spoke to his fiancée's priest.

'Appease my curiosity, if you will Father? Tell me, was Captain von
Breusch one of the officers at the war trial in Nuremberg?' Kendall's mind
paraded on a treadmill of despicable and horrific acts as described in
obituaries of Australian and Allied soldiers who suffered and died at the
hands of Nazis in that war. 'Or was he an absconder, like most of the
high-ranking officers, through Odessa? I've studied transmissions in the
field and how the enlisted men's injuries were treated with inferior drugs
due to lack of facilities, in both France and Germany.'

'It is a known fact that von Breusch had planned his escape well
before he left Berlin in forty-four, Kendall. I'm conversant with some of
the officers and their escape plans. But not of his actual desertion from
duty. In Germany's archives there should still be records of what occurred
in greater Germany. That is all I can tell ya, me boy.'

Totally absorbed in this tale of intrigue, Kendall had forgotten the
time, until Brianna nudged his elbow. Instead of arguing with his fiancée
he frowned at her. Father George had a fair idea of what she wanted.

'I mustn't detain ya good folks. Let me tell ya this, some say he
suicided. It seemed unbelievable ta me. He was a cruel individual and he
would never take his own life. I did hear the Jewish Mossad were hunting
all through Europe for von Breusch. Someone had seen him in Stratford-
on-Avon and recognised him. It's all documented ya know. That was
twenty or so years gone now. I *know* he's out there somewhere, waitin for
a noose to tighten on his neck. Mark me words, laddie they'll do it yet.
To be sure they will.'

'Excuse me Father. My son will keep you talking all night.' Annoyed with Kendall's constant nattering, Gigi's tone grew sterner. 'The food is ready to be served. Helen and Harriet are trying to keep everything hot. Kirsten will be in a flap if we don't take our place at this table. I'll tell her you're ready to help Father George to his seat. I'm going to select some slices of their delicious meats and roast vegies, while you and Brianna wash your hands. Don't be long, Kendall. All your guests are waiting for you both to be seated.'

Kendall noticed his mother's face looked flushed. Holding her arm he queried, 'Are you feeling alright, Mum. You look ghastly ...'

She interjected, 'I do feel a bit tired. Nothing's wrong, just go and let me eat my meal. Brianna is sitting next to me and you are next to Kirsten. Men and women in order around a dining table is the correct decorum. I've told you that before at our evening soirees, son.' Her attention was drawn to a group milling around her future daughter-in-law. With a distinguishing gait she past them to a chair just vacated by one of the musicians.

Kendall appraised her unstable strut from the cake table cloaked in a white cloth. *Mum enjoys parading around like the queen of Sheba in all her regalia. That damn crank wallowed in making her look a fool in front of Brianna whom he only met for a split second in our hall. Well, he can wallow in his own damn misery. The drunken bum swills cheap whiskey from dawn to dusk. I detested him most when he cocked his thumb at me in a disgusting gesture. His eyes oozed hate as his clenched fist ploughed through the air in anger. Thank God he won't be here tonight.*

Gigi eased her son's vacated chair back a little and sat beside Brianna. Kendall was standing two feet away talking to her priest and his friend Keith. 'I overheard him talking to someone on the phone as I entered his den just before we left home. I have a feeling that Ian then dialled 199.'

'Isn't that number technicians dial to test a line, Kendall? I buzzed it myself yesterday, until they fixed my party line. You dial 199 and then put the receiver down. If your line's okay, the phone rings loud and clear. It's a terrific system and works every time. Then an observant listener thinks there's a person on the line talking to you.'

'Kendall, what exactly did your father say tonight?' asked the Canberra senator who'd just joined his friends. 'Your mother, whom

I met for the first time this evening, seems quite perturbed and rather confused...'

'We can't speak here with all this confusion, Senator Bucknell. Let's adjourn to the breezeway. The rain's stopped and it'll be cooler out there. Nobody will overhear us talking. He said I was conceived out of wedlock and a huge jeer hissed from his lips. The pig accused my *innocent* mother of cajoling her to marrying him, to save my name. I've known since childhood that he's hated me. Tonight, at home he accused me of persuading my mother to turn against him. Which was a load of hogwash. What's more, he said that I made his life absolute hell. I think of him as scum of the earth.' Turning his head in all directions Kendall's mouth distorted with the pain of remembrance. After a breather, he continued with his rancorous, though moderate outburst. 'He even accused me of ruining his marriage and his life. As usual he'll probably be inebriated. I don't care if the sod drops dead before morning.' Still furious, Kendall clewed his fist into a tight ball.

'I'm genuinely sorry to have made you relive that shocking incident, Kendall. I was forewarned by Bjorn. I needed to hear it from you, in case he goes berserk and does something rash at home. If you don't object, I will drive your mother over to Abergeldie tonight. This is my personal card. If you need anything urgent, call me. I'm home for a week. My time's flexible.'

Holding his restricted throat Kendall couldn't answer, until a youthful voice he recognised echoed through the breezeway. 'Excuse me butting in, Doctor Ross. My brother Corey has a ten-week-old pup in his car. The breeder put his papers in with several leads, and instructions. They're with a basket of cooked meals under a cloth in *his* basket. This is her phone number and address. Corey bought the red setter for your mum. It's her Christmas present.'

'Everyone's in the marquee. You can join in the festivities, lad. I'll tell your brother to bring the pup into the house. Don't say a word to Mum. Let's give her a surprise.'

The teenager nodded and took off to do Kendall's bidding, with a huge grin.

'Benji Sidle's a bonza kid, and he adores his two brothers. By the size of his grin, I'd say Corey won his bout tonight. I've known their parents for ten or so years, Kendall. The boys never alter. They're always pleasant and they respect their elders.'

'Peter, those boys are the life blood of Yass. They certainly keep the community on their toes. Corey idolises Mum and she does him. Bjorn reckons those three youngsters keep him amused with their antics, even on a grey winter's day.'

Senator Bucknell waited in the breezeway for Kendall and Corey, who carried the pup. Then he followed them indoors. Placing the pup's basket, loaded with goodies, near the lounge doors, he acknowledged Kendall's signal to remain still.

'Close your eyes Mum, and hold out both hands. Don't ask questions either.'

'What are you doing, Kendall?' She frowned. 'Bjorn's foot is aching, and so are mine. Kirsten and I came in here to have a quiet cup of tea. I'll do as you've asked. But make it snappy. I need to drink mine while it's hot.' She set the cup and saucer on a side table. 'I can't keep my eyes closed for long.' She gasped as the wriggling pup settled on her palms.

Caressing the fluffy bundle, a smile appeared even before her eyes opened. A damp nose touched her chin as the red setter licked Gigi's cheek. 'Kendall, what a beautiful puppy. You darling boy, you have given me the best present I could ever wish for. Corey, she's adorable. What's her name, lad?'

'Her kennel name is Loyal Sovereign. My brothers and I call her Goldie. We thought it suited her gentle personality and glossy coat of deep amber.'

'After your spontaneous description of her colouring, I shall call her Amber.' Her dreamy eyes peered up at Gigi and she whimpered to be cuddled. 'Is she toilet trained? And how will I know if she needs a run in the yard?'

'You'll feel her nudge your knee, or lick it. She won't squat or piddle indoors. She sleeps in the basket and chews on this stick of specially-treated cowhide. I bought it from our local vet yesterday. It's quite safe. She won't choke. And she loves a brisk walk every morning when it's cool or before the sun rises, Mrs Ross.'

Gigi thanked Corey and returned to the marquee. 'Well, it *is* way past my bedtime. Kirsten, would you and Bjorn mind if I said goodnight.

I won't disturb my son, who's talking to that polite gentleman. Brianna looks a little bored with these birthday festivities. Perhaps, she may prefer to come home with me?'

'That polite gentleman is Senator Bucknell. After Brianna and Kendall have cut their cakes, he offered to drive you over to Abergeldie. I'll tell Peter and your son now. Bjorn has asked Helen to fetch your coat and things from our bedroom. Gigi, please don't let Ian upset you again tonight. If you need either of us anytime, give me a buzz.'

Swallowing, her throat was congealed with mucus. *I wonder if Ian's in another of his drunken stupors. Then his mind gets so befuddled and he fails to communicate with me. He's probably drunk himself to sleep by now.*

Gigi looked at the broach watch on her lapel. 'It's gone eleven. Where has the time flown? I imagined it to be ten. This headache of mine is accelerating. He caused it and I can't think clearly, Kirsten.'

'You'll be fine after a good rest. Darling I'll see you in the morning.'

Ready to leave for home, Kendall collected his suitcoat from the back of Brianna's chair.

'Come on darling. Let's get moving or I *will* fall asleep. Mum is ready to leave with Senator Bucknell. Once my car is packed with our gifts, I want to ask Helen for a slice of both our cakes. Then we can follow his car home. It has been a terrific night and I'm sure everyone enjoyed themselves. With most of our friends on their way home, I'll be glad to snuggle down in a warm bed. The rain's gone, but this breeze is quite cool.'

'Thanks for footing the bill for tonight's shindig, darls.' Bjorn yawned. 'It looks as though Kendall and his fiancée enjoyed the tom-toms and modern music. Gigi looks a little fragile. I guess she's still reeling from Ian's harsh rebuke. If he doesn't toe the mark when they go home, he ought to be shot. He said something to me the other night about leaving for parts unknown. I bet he has Melbourne or Sydney in his sites. I wouldn't put it past him to leave Abergeldie for good. It couldn't come quick enough for me. I've never hated anyone in my life. Boy, oh boy do I hate him after what he did in front of their guests.'

'What do you mean Bjorn,' his wife asked undressing for bed.

'Kendall told me, Ian paraded around half naked in front of their guests. Brianna was shocked and her priest seemed dumbfounded. Although neither actually saw Ian, they both heard him ranting like a wild bull. It occurred just before they left to come over here. Kendall apparently told his father that he didn't appreciate seeing him naked. It disgusted him. Their eye contact became frozen. It remained stable between father and son for some seconds, so he reckoned. Ian flinched and lowered his eyes and glared at the floor.'

'I don't feel like sleeping. It's so damned hot and oppressive. I might read for a while on the front verandah. It will be a lot cooler out there, Bjorn.' She made a pot of fresh tea and carried it and the daily paper up the hall.

Halfway up Jalna's drive Brianna gasped. 'Kendall, your mother's forgotten the pup.' Brianna looked at the rear seat. 'His basket stuffed with his things and the food hamper are behind you.'

'Let me concentrate, there's traffic ahead. Could be an accident.' The senator's car turned in through Abergeldie's gates in front of Kendall. 'Sorry, what did you say about the pup, Brianna? Now I remember. He's in the car in front of us with Father George and my mother. Bucknell just waved us on. Something must've happened. I can't stop on this curve. We'll see them down at the house. It can't have been a serious accident.'

Kendall pulled his car to a halt beside the steps and frowned. 'Why is that damn front door of ours open? I shut it myself at eight.' Mystified, he alighted and walked around to hold the car door open for Brianna. 'Darling, wait here please. A burglar or someone with evil intent could be ransacking the house.'

'Your father may have opened the door to let some air flow through the house.'

'No Bridy. He never leaves any door undone, if he's home. It looks a bit suspicious to me,' Kendall confirmed with a look of dismay. 'I'll be back in a minute.' It took him seven to go right thought every room in the homestead. 'The bird has flown. All he left is this letter for Mum. Here they are now. I won't give it her tonight. It can wait until the morning...'

'Give me what, Kendall. What's that letter in your hand?'

'Don't panic Mother. There's nothing wrong. If you support Father George by his arm, he'll manage these steps. I'll carry in your pup and her basket.' Kendall invited Senator Bucknell to stay for coffee. 'Go on through to the lounge, Peter. Harriet's just arrived, and she'll make it.

You're driving, so no alcoholic drinks? Would you prefer a cup of Earl Grey or ordinary tea?'

'Black, weak tea and no sugar will be fine, thanks Kendall. Where will I put your mum's pup? She went to sleep in the priest's arms and I rescued her. The puppy I mean. This red setter is well groomed and her curly pelt has a beautiful, glossy shine. You can always tell a champion setter by the curved peak on the crown of its forehead. I was a breeder in London for decades, long before I signed on with the Diplomatic Corps in Canberra.'

Father Brady sat and swigged a nip of whiskey while Senator Bucknell talked to Kendall in Abergeldie's kitchen. The urge to show them what he'd hidden in his missal was growing stronger with every breath.

'Next time, I'll forego my Hippocratic Oath and finish my old man. He came off lightly this time. You have no idea Peter how he mistreats my mother. I feel resentful and it takes all my energy to curb the idea of shooting him. This I found on the floor behind his desk in the den. It looks like a stub from a house of ill repute. Although it's torn,' Kendall pointed to card, 'the icon looks like The Red Dragon, but I can't read its full title.'

'Have you heard your father mention anything of this eatery? To me, it sounds as if it could be a restaurant at Moonee Ponds in Melbourne. The other alternative is it might be a sleazy joint at China Town in Sydney. Both areas have long histories of drug-dealers and pimps roaming the streets to gain a sale from some stupid kid under the influence. I can and will check on finding this place. You're not going anywhere in the near future, are you Kendall? Every query I ask has a specific reason. I speak straight from the heart. No dithering or digging for answers. Most times the person being interviewed either lies or conceals the truth, to hoodwink or confuse the interrogator. Shooting straight from the rule book, I find their replies linger on the borderline of truth.'

11.30 pm. Unable to contact Abergeldie by phone, Bjorn had driven over to see why. He sensed something was wrong with Gigi. On approaching the front steps, he heard voices raised in anger. 'Struth, I hope Ian's not wielding that damn strap again.'

'Read the letter Mum, before you go off half-cocked. It says he's gone to Canberra. And it seems his direct flight to London left four hours ago. His aged mother is supposed to be dying in Glasgow. But I doubt if it's the truth. You know he's a liar. The bastard's insane. Stephen Jarvis agrees with me…the maniac is certifiable.'

Collecting his coat, Kendall turned to the priest, who like everyone had heard their original tiff. 'You promised to show me what's in your missal. Father, I'm going to my room, can I fetch it for you?'

Brianna was tempted to interject and hesitated. 'Darling, while you're there, will you bring in our presents from the car. We can't afford to leave them overnight. I will need my nasal spray and my purse. There's a card in it for your mother.' Resting a hand on his, her genuine concern showed in Brianna's eyes for his bruised knuckles.

Relieved and thankful that tonight was drawing to a close, Kendall tossed his sweat-soaked shirt on the bed and selected one suitable to wear on such an oppressive evening. Trying to balance the parcels and Brianna's purse he found the steps a challenge in the dark. 'Damn this front light, the bulb has fused. I better replace it before someone has an accident or a fall on these wooden steps. This lower one needs anchoring to the wall.'

Clad in a cool shirt and casual shorts, Kendall tried to open the door with his elbow. Brianna came to his aid, carrying a gift for his birthday. Letting the wire door swing of its own accord, he nudged the heavy, main wooden door back to prevent dropping an armful of gifts.

'Darling, put the top parcels on your bed. I can manage the rest.' He blew her a kiss. 'This lot are my mother's things. I'm damned if I know why she wanted a pup. With five border collies and three piebald red kelpies we have enough dogs to sink a battleship on this property. The red setter is a bitch and in time she may bond with Mum.'

Treading Abergeldie's floorboards Bjorn's mind lingered in a field of dismay. *I find it difficult to believe, why that imbecile implied Kendall's mother was a whore. What a despicable thing to say about his wife. Ian's a no-hoper. And an idiot to keep threatening Gigi like he does. I admire Kendall for thumping him. Ian deserved a good hiding. The scum-bag's full of wind. Why didn't I shoot him months ago in his den when he refused to sign everything over to Gigi? Bucknell will pull the bastard into line. Kendall's mum has drummed into him the importance of being honest and truthful.*

Bjorn knuckled the wood of their front door as Kendall walked up the main hall. Exercising his vocal cords, he asked sarcastically, 'What happened to your phone, BJ? I thought we'd seen the last of you for one night.'

'My telephone works okay. Yours *doesn't*. Has your old fart left it off the hook? I've been trying to raise Abergeldie for half an hour. Kendall, is your mum okay? She forgot her purse. In there is a bottle of angina tablets inside it. My wife found her bag under our bed. Kirsten's in the car. We were concerned, in case Ian was up to his old tricks.'

'Sorry I snarled at you, Bjorn. Probably the people on our party line have left their receiver off the hook. In ten minutes I'll buzz them on our special button. If they don't respond, I'll be stuffed until they do answer. Tomorrow I'll re-apply for a *silent number* and a *new* line. If the PMG refuse to grant my third request, I *will* contact their board of directors in Sydney and demand a separate phone line.'

26

On the plane Ian Ross looked at his injured nose. His image reflected in the shaded window showed minimal damage. Dressed in a conservative suit of navy he unknotted his silver and blue-swirled tie. Refolding his shirt collar down he proceeded to comb his hair in a different style. His next task was to activate the escape plan and a new destination.

Before packing his suitcases, Ian had rummaged through his wife's wall safe and he removed most of her valuable jewellery. Unbeknown to her Ian was conversant with its new combination. In her office he'd watched a local mechanic resetting the tumblers, due to a previous argument over the property's finances. Later in the day he had observed Gigi checking each number as it clicked into place.

Having depleted most of her important documents at noon and having pilfered rolls of large denominational notes, he then reset the tumblers, and replaced the landscape. This concealed the wall safe. Gigi would be none the wiser when she came home from the Yass hair salon.

Ian had selected only the genuine pieces of his first wife's jewellery from that safe. Erika's emerald necklace with its distinctive design could be readily traced. Somewhere in his wardrobe were two photographs of Gigi wearing his deceased wife's engagement ring and earrings. Other items of jewellery and pouches containing rare gems, plus her pearls were also concealed in the hidden pockets of his tan satchel, along with several diamond bracelets, gifts from Pierre Jean Paul Bouvier, in France.

Relaxing in transit, Ian methodically leafed through and viewed every document in his tan briefcase. Unnecessary documents were still in his desk in Yass. Ian then proceeded to forage through all the data to find his current passport. The third forged passport, under a fictitious name was safe in his suitcoat pocket. His personal documents were in the larger of his two cases, secured somewhere in the aircraft's hold, six or more feet beneath his airline seat. *Everything packed in these hidden pockets of my valise will be undetectable, if Customs search it for the third time, on my arrival in Sydney in ten minutes. It should be easy to hail a taxi to a hotel. I must remember to update my photographs in these forged passports. The photo in this current one will be outmoded, once I dye this hair of mine and change my appearance. My agent in Kings Cross will know whom to contact regarding this problem. All will be well within the hour.*

Every bit of data needed for this flight to freedom was stashed in his eagle-crested tan valise. This satchel could be reorganised at the last moment, if necessary. Ian had salvaged extremely important documents from the locked desk drawer in his den. One of the items was a sealed packet of a hair dye and a false moustache, hidden there for just such an emergency.

The moment he exited Custom's barrier Ian walked to a telephone booth and dialled two unlisted phone numbers to make appointments for the following day. Neither of his overseas flights could be finalised until he contacted Mascot's International Terminal to secure a booking. Apart from him, only the airport personnel would be aware of his new identity and his altered destination.

The last phone call left him free to find a secluded eatery for his sojourn in Sydney. Choosing the correct documents needed for a week's stay in this city couldn't be rushed. Even the smallest detail must be thoroughly evaluated, before he attempted to meld in with crowds of business people on their way to work or visitors sightseeing. His disappearance into a void of the masses must be planned to perfection. By collating each item in his valise, it allowed Ian to select the illegal documents needed to pass muster at various hotels.

In the privacy of his room the hidden pockets of his valise were emptied and neatly restacked with the required data. These included rolls of Australian bank notes and the jewellery confiscated from Gigi's wall safe. The three forged passports remained under the false base of this satchel, in their original position.

Ian had packed enough clothes for a month. Or until he could purchase fashionable clothes from The House of Distinction, an exclusive men's shop he often frequented in Balmain, a suburb of Sydney. The note he had left at home would undoubtedly work as a red herring. It falsified his real destination and would be bum-steer to Federal investigators who, in time, would search every inch of Abergeldie's homestead in Yass. A quick reschedule of his preplanned journey took an hour while imbibing on coffee in a small café in Clarence Street.

Paying the bill, he picked up a Canadian brochure and thought, *Tomorrow I must take a short walk to the General Post Office [GPO, he jotted down in his diary] to send a cablegram to both my banking houses in Switzerland, and Luxenberg. They have never contacted, or sent me a statement for more than a decade. I've tried to contact them numerous times. Either my letters have gone astray, or they ended up in some dead-letter office on the continent.* Ian then strode fifty yards to his hotel, re-collected the room key and retired.

6am Next morning. Tinting his hair to mid-brown he rolled everything in a towel, rinsed the shower and, plotting each future move carefully, then wrote a note to his overseas contacts. On his walk to the car, Ian threw his room keys on the hotel desk along with ample money to cover his twenty-four hour stay. Beside the phone he paused to search the outer pockets of his briefcase.

Before reaching the taxi rank on his way to Kingsford Smith International Terminal, he disposed of the empty tube of dye and towel in a bin. At the airport he secured cancellation on an international carrier for that afternoon's flight he booked only a small case full of current newspapers and a curler-bag he found in the toilet, to Los Angles in America. With time to spare, he left to have lunch in a small café.

By the time everyone and the Feds realise I'm no longer in Australia, they'll presume I'm on a plane to America. Or that I may've stopped off in Hawaii. My only regret is not having seduced Brianna O'Shea when the opportunity arose last evening. Still, part of my revenge is having pilfered my wife's precious trinkets and the valuable oil A Nude in Love. They'll serve me well in this, my escape to freedom.

11.39 pm the previous evening. Festivities were forgotten on Abergeldie and ready to retire Father Brady removed a photograph, no bigger than an average postage stamp, from the back of his missal and showed it Kendall. A paper hidden in its spine once belonged to the original owner of this miniature. Nobody gave Ian Ross one iota of thought, least of all his son Kendall.

Both he and Brianna had enjoyed the evening amongst cherished friends many of whom had travelled from interstate and overseas to celebrate their engagement.

In a constant flurry regarding her husband's mysterious disappearance, Gigi found it difficult to settle in her room. She refrained from disclosing how miserable she felt over the bitter things Ian had implied.

She refused to let their argument to spoil her son and Brianna's first evening at home. Or what was left of it. They deserved to have more pleasant memories of the night on the cusp of fading into oblivion; especially after the debacle earlier this evening.

The adorable red setter pup licked Gigi's fingers as she settled it on a pillow in the wicker basket.

'You can sleep in my room tonight, Amber. Then, tomorrow everywhere I go you will be with me. After our daily walks, I'll brush your coat to keep it glossy and in good condition, free of burs. Kendall will make a bed for you in my office. I see you've found the bowl of clean water he put on this floor. You are a good girl. You haven't whinged or whined all night. You really are a darling puppy, and I love you, Amber.'

'Mum, you were looking forward to our party for so long and he spoilt it. I'm pleased he's gone, perhaps for good. Not one of our guests are feeling miserable. You're free to lead your life, without a grumpy bum nitpicking everything you do. Look at it as providence. To be truthful, I won't miss him. The bastard couldn't keep his nose out of my personal business and yours. His ugly face will always be a prominent feature in my mind. I will never forget him, or forgive his cruelty to you.'

'On a lighter note, Kendall he's done us all a huge favour by leaving tonight. Father George is ready to show us the miniature photo. Come and sit by me on the lounge.' Gigi gave its tapestry a pat. Harriet will be our personal waiter. Tea for me thanks dear.'

'What do you prefer Kendall? Maybe a slice of Miss Brianna 21st birthday cake, or your birthday one topped with whipped cream or ice-cream?'

'Lemon ice-cream on a slice of chocolate cake sounds delicious. Thanks Harriet.'

'Kendall, is Brianna okay? I noticed tears in her eyes on her arrival home.'

'She cried when I untied my birthday and Christmas present. Look what she bought me in Dubai. It is eighteen carat gold, Mum. I thought it was a silver wristwatch.'

'One cup of hot tea coming up. I sent Rose to bed a minute ago. She and Jasmine served your guests their midnight refreshments. Now you relax and enjoy yours, Gigi.'

'You must agree with me that the party was a splendid success without Ian's nasty innuendos disturbing the peace.'

Hardly knowing what to say or where to look, Gigi was flabbergasted by Kirsten's comment. 'Well, I suppose you're right,' she agreed, yet still felt apprehensive. 'If I feel unsettled I might let the pup sleep on my bed tonight. I'm awfully jaded and I really feel drained after that rumpus earlier. Oh, it's hard to believe why Ian would say such horrible things about me. The lies weren't true, you know that Kirsty.'

'I am conversant with the truth. Those fantasies were just a figment of your husband's sordid imagination. Ideas he's conjured up because of his insane jealousy of Kendall. Nobody needs to convince me, how Ian's mind works. I suspected what he might be up to, long before he absconded.'

'You must believe me Kirsty, I never forced him into marriage. Quite the contrary, I told Ian to leave me in Gloucester. I absolutely refused to have Kendall aborted.' Gigi swallowed hard in an effort to prevent tears forming in her eyes. Yet they still ebbed free.

Kirsty tried to convince Gigi that none of this was her fault. 'Gigi, can't you understand why Ian manipulated your mind? He was devious and he purposely instilled his conniving ideas into your head to make you feel guilty. His evil ideas didn't wash with me.' Kirsty held her neighbour and best friend to her breast in a comforting embrace. This gave Gigi courage to weep freely without Kendall seeing or realising her anguish.

From experience, Kirsty knew how Ian Ross had behaved with other women when his wife was in hospital or on business trips. *What made a beast like him behave so uncouthly? Gigi shouldn't reproach herself over his infidelity. Ian wasn't worth anyone's pity. I'm glad he can't harm her any more. God knows where's he's gone. I don't believe his mother is ill in*

Scotland. He could be on a plane to hell for all I care. That pathetic and lustful creature deserved to be in hell over way he mistreated Gigi.

Passing her a handkerchief, she tried to hide her own dismay.

'Come on you silly duffer, dry those tears. Everyone's anxious to try those delicious cakes. Neither your son nor Brianna will attempt to eat theirs, until you smile. Gigi, forget Ian and come and see Father's photograph. He and Kendall are talking to Stephen and Senator Bucknell. Peter's witty, dry sense of humour will put a sparkle in your eyes. Bjorn and I are quite fond of Peter. His secretary, Deidre Sanderson couldn't accompany him to the party. Dee is busy catching up with his business letters this weekend. I think he missed her a lot tonight. Oops last night. It's one now. I think we ought to make tracks for home soon. It was a wonderful evening, despite the rain and Ian's absence proved to be a blessing.'

Drying her eyes, Gigi powdered her flushed cheeks, put on fresh lipstick then wanted to look at the priest's miniature photograph and listen to his narration of a letter. Serenely walking across the lounge room, even though she felt unreceptive to idle chatter, she tried to be affable. Beneath her charming façade lay an anguished heart burning with the fire of betrayal.

Feeling weary after chatting to the diplomat and his friends for what seemed ages, Kendall knew it would be hard for his mother to dismiss the earlier incident of his father's nudity and brutality on Abergeldie. However, he preferred to forget everything.

'Mum, forget him and sit here until Father George has finished his whiskey. What did you think of my new watch? Pretty snazzy and Brianna loved the cultured pearls I bought her in London. Originally, they belonged to an elderly doctor, who sold them to retire in Coventry. Bridy chose the pearls, and has the guarantee and a certificate of accreditation.'

'You received some exquisite gifts for your engagement. Tomorrow you can show them to me in daylight. I only got a peep at some of the unwrapped presents, with everyone milling close to the cake table. Son, I am sorry over the argument with your father. I now admit that Ian deserved a hard thump in the nose.'

'Mum, the only regret I have is, that I should have done it years ago. He'll survive, unfortunately. I'll be fine in the morning. Oh, today.'

Kendall shook his head. He wanted his mother to face the truth and be honest with herself. 'It hurt me, not physically but emotionally, to hear him denigrate you in front of strangers. And for that, I shall never forgive the beastly man.'

'While I'm waiting to see the miniature photo, I'll take Amber out for toilet walk, Kendall. Then she should sleep until breakfast time.'

'Righto Mum. Unleash her on the verandah and she'll find a spot to wee. A short whistle and she'll come back to you. Do not walk down those steps on your own. I'll come and fetch you and the pup. First, I need to answer my call of nature urgently.'

Swirling her head to the rhythm under a star-studded sky, the lilting tune triggered her mind into recall of her younger days in Llangollen, on the last visit to Manny's doctor. The news of her pregnancy seemed fifty decades away now. *I'm so proud of my son. He weathered the tempests of hate and disownment by an arrogant father, who denied his existence. Kendall won the battle and he can now face life free of that tyrant's domination.* Not allowed to sing or hum in Ian's presence, her mind created orchestral rhythms and tunes sweeter than a lark in springtime as nature's music echoed across their paddocks.

Re-entering the house Amber nuzzled her knee. Gigi led the pup down their main hall. In the lounge room she settled on a mat in between her feet, until Kendall beckoned his mother to remain seated.

Kendall grimaced and moved back a little. 'Father George please don't let my Mother see this miniature. She will collapse with fear and remorse. I never suspected him to be a Nazi officer who worked in the Berlin Chancellery. Did von Breusch really order your death and sign the death warrant? That's downright murder.'

'Captain von Breusch didn't know that me friend Father Patrick Kelly went to face his firing squad instead of me. Patrick gave up his life to save mine. I did hear a loud volley of shots as I shuffled down the Chancellery's front steps. It took a long time before I could move to Den Linden Platz. Young boys were playing in the rubble of bombed buildings. I remember they huddled together in a doorway as I approached them. With tears in me eyes, I tried ta help the lads, but I couldn't. Not long after, I fell on the wet pavement, Kendall. Then I clung on ta the bough of a linden tree, until a black-tinted car stopped to collect me on that corner.'

Gigi ducked under her son's arm to get a peep at the priest's photo. One look and she exclaimed, 'Oh my God, Ian was a Nazi officer.' Stephen Jarvis caught her as she fainted. He and Peter Bucknell carried Gigi to her room.

'Until Doctor Jarvis has accessed my Mother's condition, please stay here and don't attempt to move Father George. This incident has shocked you immensely. I'll bring you something to settle your heart palpitations. Mr Svensson will stay with you until I return.'

'It's been a long, stressful night and even a longer day for you, Father,' said Bjorn. 'My wife is getting a cool compress to ease your head. I don't think it's advisable for you to sip that whiskey. Once the initial shock passes, you will feel less shaky.'

On his return, Kendall found Brianna trying to console her priest, who was rocking back and forth as he moaned.

'What have I done? I should've burnt the miniature of von Breusch long ago. I didn't mean ta upset the dear lady. She fainted and I caused her to collapse. Kendall, I am the criminal here, not you, me boy. How is ya mother now? I hope in time you both can find it in ya hearts ta forgive an old fool. I am ta blame for me stupidity.'

This oppressive heat gradually began to subside as the zephyr grew stronger. Bjorn decided to stretch his legs on Abergeldie's front verandah. 'A good drenching will cool this stifling hot and dust-laden air.' Shifting one foot off the back rail he rendered a sigh of tiredness. It mingled with a cry of the tawny owl sitting on a branch partly hidden from his view by leaves of the massive, cedar tree. Cone-wooden roses littered the dry grass under its spreading boughs.

'I feel utterly stitched-up in this damn heat and the weather's so oppressive. Peter, did you manage to contact your boss in Canberra? The mystery of this Ross saga is growing into a frustrating saga. God knows where Ian is and it's anyone's guess. I reckon the mongrel's gone to ground in Sydney, or in Melbourne.'

Senator Bucknell scratched the mosquito bite on his index knuckle.

'Yes, Bjorn I did. He's going to contact our Federal blokes as soon the dawn breaks. Everyone here on Abergeldie will be inundated with a team of them within a day or so. Is there a solid flat area in those top paddocks for a small aircraft or helicopter to land safely?'

Bjorn thought this sounded a bit odd. 'There's a flat plateau on the far end paddock. It'll need levelling with a grader to settle the dust. Kendall knows a local bloke who owes him a favour. Ask him to arrange something. We're heading home now. Kirsten's tired and she's just given Gigi a sponge bath. Poor darling, she didn't need another shock like this one. Fancy that bastard being an escaped Nazi. Peter, the news didn't come as a shock to me. I suspected for quite a while that Ian wasn't Scottish, or as we know him now as Rolf von Breusch. I've heard him swear several times in a dialect which I took to be Austrian in origin. Lanky Hancock can vouch that he's also heard Ian cursing numerous times, when he's been sozzled to the eye-balls and in a foreign lingo.'

'From tomorrow on, Abergeldie will be out of bounds to all strangers. Every member of the staff will be restricted to their quarters. Unless they need to go into town, then that person will be accompanied by a Federal officer. All travel will be banned, and special passes will be issued to the family only. The Feds have strict rulings and they insist on everyone abiding by them.'

27

5 am. A wind stirred the rustling leaves as Kendall awoke with a startled look on his furrowed brow.

'What made you growl, Amber? Did you hear strange footsteps in the hall outside mum's door? Let's have a look.' His finger raised. 'No barking, or you could disturb the prowler.' Kendall donned his dressing gown and gingerly unclipped the doorlatch. 'Sit here, you know that command girl.' He looked up and down the hallway. 'Come pup, there's a dark figure about to enter your mistress's room. From here I'd say it could be Didier Smith. He's a local lout. What's he doing here at this hour? Stay on guard Amber, while I tackle the thug.'

The pup snarled and bared his teeth as Kendall approached his mother's room. The door sat ajar, and snoring could be heard in the otherwise silent room. Kendall wielded his cricket bat swiftly. It caught the intruder unawares. 'I clouted you hard on the noggin, Didier. Don't try to escape, or move. If you take one step towards either my mother's bed, or Jarvis asleep in that chair this red setter will lunge. She is a trained watchdog and she does bite.'

Petrified he stammered, 'You…your old man threatened to shoot my father over a debt of money. I wasn't going to hurt your mum. I demand to speak with the bastard now. Ian won't kill my dad, because I'll bash his skull in first with this tyre-iron. Do ya want me to use it on you? I will, unless you let me go, Kendall.'

Standing behind the intruder, Bucknell's sudden swing propelled the cricket bat with precision as it collected and shattered the oval, wardrobe mirror. Snarling and with teeth bared, the pup tackled Didier and knocked him flying. The rumpus had awoken both Gigi and Doctor Jarvis.

Kendall grabbed the ruffian by the scruff of his neck. 'Get out of this house and stay off my property, you damn rabble-rouser. If I catch you anywhere near here again, your next journey will be in a locked paddy wagon to Yass Police Station. I intend to report your uncouth threats and activities to the local coppers. And you will be arrested before this day is through. Heed my warning and leave now Didier. This dog and I will escort you to the front door. Remember, on my command, she will take a huge chunk out of your bum.'

Kendall remembered Brianna. 'Struth, I hope this thug hasn't awakened her with his rough antics in Mum's room.' He watched the lout ride up Abergeldie's drive then closed and locked the wire door and checked if both their guests were okay.

Brianna, unperturbed by the noise, sat reading her mother's card on the bed. Father Brady hadn't disturbed. He continued snoring Paddy's pigs to market. With loads of scores to settle, Kendall went to his mother's room to see how she was feeling now.

'I'm fine thanks; son. But I have an urgent call of nature.' Gigi excused herself. 'I'm a bit unsteady on these shaky pins of mine. I heard Kirsten in the hall a moment ago.'

'Bjorn was talking to his wife, I think.'

'No, he isn't Kendall. I'll walk your mother to the bathroom, and make her bed with clean sheets.' Kirsten nodded to Bjorn. 'Ask Harriet to get them and a couple of fresh towels, please darls. Gigi will need a fresh nightie. There's a cotton one on the chair. Pass it to me, Bjorn. We girls need privacy, then all you men can leave the room.'

He and Jarvis, with their diplomat friend Peter Bucknell, made a hasty retreat to the main hall. Father Brady poked his nose out his door. Even though he'd seen their intruder fleeing on his horse up the drive he didn't ask Kendall why. The tyre-iron with Didier's fingerprints on its metal handle was proof of his unauthorised entry.

'That miserable lout won't show his face around here in a hurry. His father's a no-hoper, minus a brain that functions. Their whole family live on hand-outs, or steal what they need to survive. None of them know what a hard day's work means. They'd pinch the eye out of a bag-needle and come back for its shaft.'

No sooner had Kendall spoken than his mother returned from her bathroom. 'Son, will you ask Harriet or her daughter to clean my toilet floor. Have you forgotten Jalna and all our staff will be working tomorrow, preparing for the Cootamundra Show?'

'No, I haven't Mum. I'll help you over to the bed. Then Jarvis is going to give you an injection. The drug will settle your nerves and calm you. I know that I've neglected you, and I forgot to give Father George his medication. Stress does weird things to people. Take your time, the floorboards are uneven and you can't risk having a fall.'

'I'm not crippled yet and I can walk to the bed. Will your friends think me rude if I sleep for a couple of hours? I feel awfully tired son. My head's spinning and I can't afford to faint again or make a fool of myself. Leave my albums on the dining table for Brianna to browse through later.'

Kendall acknowledged his mother's polite request. He promised not to allow anyone to peek at her delicate photographs. He reminded her two of their guests Senator Bucknell and the priest were leaving tomorrow. The priest for a congressional seminar in Goulburn.

Brianna took the initiative to tell Kendall that Father George wouldn't be returning to Abergeldie for week. 'Father will appreciate relaxing to look at your mother's photos then, darling.'

'Well, you're the boss. But only until tonight.' Deliriously happy, Kendall smiled and gave her a huge hug.

Bjorn had parked his car a fair distance from the homestead's front steps, due to puddles. Kirsten indicated for her husband to brew the coffee. Abergeldie's cook, Harriet, finished making Gigi's bed, while Rose tipped a bucket of water down the toilet overflow hole.

'I fluffed up both your pillows and turned them over. Now you will feel more comfortable, Gigi. My daughter is making you a fresh jug of orange juice. The fruit harvested this season is of an outstanding quality. Rose and I made twenty jars of quince jam and two dozen of peach and persimmon this morning for the church fete. Heavens, my pot of blackberry and apple conserve will catch, if I don't go and move it over to side the hob.'

Talking to Kendall and Senator Bucknell on the front verandah, Bjorn thought the incident with their intruder unbelievable.

'What incited that young bludger to trespass on this property, matie? He's a thorough no-hoper. A good scrub with a wire brush might lift some of the grot from his scrawny body. It seemed unfeasible for him to get past this bitch. Did you tell him Saffron is a trained hound?'

'Yeah. Mum renamed the pup Amber. I reckon Saffron sounds better than Amber. Although it is the colour of her coat. It won't take the pup long to respond to her new name. Mum usually walks her to the gates or down to our shearing sheds most mornings. Now Brianna and I will take the pup for her daily constitutional, seeing Mum's not well.'

Bjorn dismissed this in favour of seeking the truth. On reflection he recalled Ian's vehicle had been missing for at least two days. Rather than commenting to Kendall, he gestured for Peter Bucknell to follow him out to Abergeldie's first cool room.

'Hey mate, I think Ian may have gone to ground in your neck of the woods. He often boasted about the young titters in Canberra's business district. I think it's more feasible than him absconding to Sydney or Melbourne. Do you think the Feds should go through his den? Kendall found this under his desk and gave it to me for safekeeping.'

'This looks like a restaurant meal chit. Fetch the key and we'll check his office. Never can tell what might turn up. No, on second thoughts we better not intrude. The Feds will demand that privilege. Their team will be here first thing tomorrow. They'll check every room in the house and its sheds. God alone knows what they'll find. Those two-legged bloodhounds will unearth the slightest thing that we could miss in a full day's search.'

'I forgot to ask Kendall if those top paddocks are accessible. The Cessna aircraft is due to land, wind permitting, at six am. There will be a full crew onboard who'll take over from me. As you know, I've been recalled to Canberra, and I'll return on that plane.'

Bjorn moved down a step to relieve his stiff knees. 'Kendall has arranged for a local bloke to bring in a grader, a water-tanker and a steamroller this arvo. That area will be flatter than the head of a tack before morning. It's a dustbowl all across our top paddocks at present. That skerrick of rain last night didn't even dampen the rye grass.'

Kendall greeted the diplomat outside his room. A flexed eyebrow coupled with a side nod signalled Bucknell to follow him down to the lounge room. There they could talk without disturbing his mother.

Kirsten pulled the chair closer to Gigi's bed.

'Harriet's making vegetable broth for your lunch, darling. You need the sustenance to recoup your strength. The shock of seeing those photos coupled with the recent incident has depleted your strength and you lack the ability to cope with more worry. I used your secateurs to cut these lemon and lilac rosebuds from the front garden. Their fragrance is delicate, not strong. I'll put the vase on your windowsill, then you can admire their beauty, Gigi.'

Minutes later Kendall handed Bucknell his mother's letter he'd just read.

'I think you should read it, Peter. It wasn't sealed so Mum and I are aware of its contents.' Turning to their neighbour he nodded. 'You were right Bjorn about his mother. He reckoned she was ill in Glasgow. All the years we've lived in Australia I've never heard him once mention my grandmother. Something doesn't gel with his excuse…'

Interrupting, Peter agreed.

'Kendall, I've already seen and photographed this letter and the priest's miniatures. This letter explains a lot and it'll save the Feds time to evaluate his movements. Every second counts after an absconder leaves home. All these items are important evidence. We cannot afford to allow a single margin of error in any case under scrutiny. I'm apologising in advance. From tomorrow on, nobody will be permitted to leave Abergeldie. All your household staff will be restricted in their movements. No outsiders will be allowed to enter this property. It will be out of bounds to all strangers and tradespeople. You will be brought up to speed by the officer in charge on his arrival. How soon can your female staff have four single bedrooms ready? The Feds will need a quiet office. And it must be accessible to a toilet and a reliable telephone.'

'No problem there. All five guest rooms are ready. They only need fresh towels. The sheets will be changed on a regular basis. Otherwise our incoming guests will eat with us in the kitchen or in their quarters. Night meals will be eaten in the dining room.'

'Will your mother be well enough for me to interview her this evening? There are a couple of issues I need to resolve, regarding your father's sudden ensconce. Minor problems that may not seem important to you. With the facts detailed in full, it will eliminate further inquiries down the track. I'll leave these matters with you, Kendall.'

'I'm glad he's absconded. He may have boarded a flight to Timbuctoo, or Suva for all I know. Pity it wasn't permanent, then we'd all have peace.' Kendall glanced at his wristwatch. 'If he ever comes home or goes near my mother again, I'll soon give him short shift. Excuse me Peter. I promised to give Brianna a tour of our stables and the saddle room. After that I can spend time taking to you and Bjorn somewhere quiet. On the front swing would be ideal. Only the wild owls or pink-breasted cockatoos will hear what I have to say out there.'

'Would it be possible to talk in your mother's office? Gigi hears everything we discuss on the front verandah.'

Kendall threw the duplicate key to Bjorn. 'Brianna and I will be back soon. Then I need to check on Mum. Otherwise her office is out of bounds to all our staff, unless she or I are working in there doing the damn accounts.'

Brianna intervened to say his mother needed him. 'Gigi's breathless and she's finding it hard to breathe. Kendall, could you or Doctor Jarvis see her now please?'

Discovering her husband was once a Nazi officer had greatly affected her. Now she was trying to forget what she'd seen. 'It can't be true. I saw him dressed in a Nazi uniform.' She kept repeating amid tears. 'He's ruined all our lives.'

This devastating news, albeit belated in context, had the potential to destroy, not only their lives, but also their close friends. They could be alienated by people whom they trusted in this small country town of Yass.

Perturbed by his mother's deteriorating condition, and her health in general, Kendall had considered hospitalising her. Jarvis disagreed with his decision.

'Look at it like I do, Kendall. Hospital is not a viable option at present. Your mother will recover from this shocking ordeal with people around her who care and have her welfare at heart.'

He nodded in agreement then considered the idea of a live-in nurse to care for his mother. 'It is inconceivable to expect any nurse doing a twenty-four stint without a break.' An exasperated gasp lacked the potential to clarify his anxiety. 'Let me think over your proposal. In time I might relent. Stephen, it wasn't easy for Mum to accept that she married a war criminal. I am the result of their union. I find it abhorrent to even think of that mongrel lying next to my mother in bed. It might be ages before she admits his brutality.'

'Her breathing and pulse are regular now, Kendall. The injection of temazepam I just gave your mother is making her drowsy. Would you or Harriet mind brewing some hot coffee, Kirsten? We could all do with a cup of *strong* hot coffee.'

Kendall nodded. 'A hot cupper will fill the gap. Nothing stronger for me, not at this time of day.'

'I agree with you, Kendall. A relieving nurse could take over at night. You can afford to pay both of them a reasonable wage. You mother will need around the clock nursing by a professional. She and you have gone through hell since receiving this devastating news. Your mother has suffered and is still suffering the torment of reality. Need I say why? Now if you will excuse me, I'm going to wash my hands and check her breathing again.'

'Excuse me, Stephen. I need to speak with Kendall alone and in private.'

They both knew it would ease the tension as Bjorn promised not to upset Gigi in case the injection hadn't worked and she wasn't asleep.

The diplomat, who had just arrived on the scene, heard what Bjorn proposed. 'If you don't mind I can perhaps be of assistance in some way, Kendall. I am at a loss to understand this problem. I'll bring your hot coffees out to you in the breezeway. Kirsten is going to tidy your mother's room and I need a spell after travelling up from Canberra.'

Peter Bucknell acknowledged Jarvis with the normal greeting. Jarvis glanced at her partly closed door and smiled. 'The injection I gave your mother will allow her to rest for some hours. I'll be here until she awakens, Kendall. I have allocated the time to stay overnight, if necessary.'

'That's fine by me. Mum needs to be bossed by a man who cares, and who ignores her obstinacy. She refuses to take her prescribed medications from me, Stephen. I'm off to sip my coffee before it gets cold. I have given Father Brady his tablets and advised him to stay in bed until his morning tea is ready.'

A smile came in the form of his answer. Nodding to Kirsten carrying a tray, it more or less indicated that Jarvis intended to see Gigi in a professional capacity.

Enjoying the vision splendour of a glorious summer sky, Bjorn challenged Kendall before he had a chance to speak.

'Hey mate, where did you get the idea that Ian might've gone to Timbuctoo? Do you know something more than you let on to me? Let's talk to Bucknell out in the breezeway, then we won't disturb your mother.'

'Okay. Timbuctoo was the first thing I could think of Bjorn. A furphy to keep him guessing. A fallacy. This morning I found another chit in the den. Read the scribble on this torn piece of a letterhead. It *is* self-explanatory.'

Bucknell gasped in disbelief. 'The print has faded a lot. If I'm not mistaken it reads as a Nazi insignia or crest of some description. What do you think it means, Kendall?'

'Peter, this is identical to the letterhead I saw inscribed on a document in his briefcase. As a kid of eight, I sneaked in to spy on the work inside in a manila folder on his desk in Evesham. We lived there before immigrating to Australia.' Kendall vaguely remembered Bjorn and Kirsten having sponsored their family to migrate.

'Hey matie, you reckon this torn fragment is identical to the one you saw back home in England, in or around the late forties?'

Kendall tucked the scrap back in his shirt pocket. 'Yes. Both these stubs I also found in his den look as if they came from the same nightclub Bjorn. I examined them under a magnifying glass and they are identical.'

This dragged the mystery of printed stubs down to a quagmire of indecisive thinking. 'I am right in suspecting they came from an irreputable house of illegal gambling. The first word as you can see is indistinct. It could be Golden or Mandarin Red Dragon. The last word is partly obliterated. Chinese in origin, would be my guess.'

'This is no surprise to me, Kendall. I've often thought Ian frequented brothels in Sydney. Perhaps it was an opium den of some kind? There are lots of well-known low-dives and brothels in Kings Cross. Chinatown harbours an element of undesirables in their gambling houses. If one can believe what the tabloids inscribe to be the truth,' Bjorn sniggered as Doctor Jarvis joined them and placed his iced coffee on the breezeway table.

'Am I intruding, Kendall? If you gentlemen were discussing something important, I don't mind drinking this indoors.'

'It's okay Stephen. I was showing them what I'd found in *his* den. There were also condoms in the drawer. I destroyed them and the box. I never use, or choose to use rubbers.'

Bjorn checked the time. 'Struth, I forgot to remind Kirsten. We're expecting my uncle to buzz us around now at home. You don't mind if I give him a quick call from your office, Kendall? Sam and Edna should be leaving Young around noon.'

'Why ask! The front office is fine. Shut the door. I don't want the pup to go in there. She's grown accustomed to squatting on the floor. Rose has just taken her for a walk. I agree with mum. She changed her name to Amber. Oh, she loves the pup.'

'Why did you mention the condoms to Bjorn? Surely he wouldn't use them?'

'No, of course not,' Kendall laughed. 'If Mum had found the box in that bastard's den she would've had a fit. She'd know they weren't mine.' Kendall paused to embrace the fresh country air. He and Stephen followed the diplomat indoors. 'It's incredible how she hasn't contracted a contagious disease from that lazy bludger. You both know, that's more that feasible with his despicable habit of sleeping with prostitutes.'

I'll take a phial of his mother's blood the moment she awakens. A full blood count will detect if she has contracted a sexually transmitted disease. Kendall will expect nothing less of me. I dislike deceiving him. But it will eliminate further doubts and we'll know the results within twenty-four hours.

'Do you realise Ian's done you both a favour by leaving. At least your mother will suffer less harassment and be sexually safe. He treated her like a street whore and worse still, he *has* considered her as one for longer than a decade, Kendall.'

Kirsten waited until the men had finished speaking. 'Brianna is waiting to unwrap her birthday gifts from Ireland and your engagement presents. She's in her room, Kendall.'

'We were on our way in, thanks Kirsten. Tell her I'll be there, as soon as I've washed my hands. You go on through to the lounge, Stephen. We'll join you in a minute.'

A short period lapsed between unwrapping the gifts until he and Brianna chose to sit on the front swing. They discussed their departures while watching the sunset change from crimson to pinkish-grey then fade to a soft blue. Kendall conveyed how concerned he was over his mother's health. The topic then switched to their wedding plans.

6.50 pm. Jarvis entered the bedroom to give Gigi an injection and heard her groaning in her sleep. 'She must be dreaming of Ian, probably of his brutality. I won't disturb her. This B12 injection isn't important. Kendall or I can administer it to her tonight.'

Meanwhile Kirsten had asked Harriet to brew an abundance of coffee and let it cool. The workmen and shearers were exhausted from slaving under a hot sun and in the oppressive heat, and would appreciate the coffee before heading for showers in their bunkhouses. The casual rouseabouts also enjoyed their cool drinks before leaving Abergeldie for their local homes.

7.30 pm. The Svenssons returned to Abergeldie with a drowsy and over-indulged priest dozing on the rear seat of their car.

'What time did the coach drop you off in Yass, Father? And did you enjoy the weekend seminar? You look as though the long hours of studying have stressed you. I've never been to one. Nor do I have any idea what the study entails.'

'The Goulburn bus dropped me off half an hour ago, Kirsten. I found the long hours of study quite strenuous at times.' He sighed and then yawned.

'Quite a lot has occurred in your absence, Father George. Relax, we should be home in ten minutes. Everyone on Abergeldie will be anxious to hear your news.' Kirsten looked around and saw Father Brady was asleep.

8.pm. The diplomat had also returned from Canberra with interesting news and answers to puzzling questions, yet with no knowledge of how to solve their dilemma. Tomorrow he and Stephen Jarvis would discuss with Kendall what they suspected might've been the cause for his father's sudden departure.

4 am. The following day: Cocooned in the blissful peace of a new dawn all those on Abergeldie slept until an awakening sun tried to break through thunderous clouds. Rain drifted in at five and caressed the parched earth with its soft coolness.

Silent and swathed in mist, Jalna's occupants began to stir around the same time to prepare for the Cootamundra Show. Helen Raddick and her daughter prayed for the rain to ease. If not, the royal icing decorations on their fruitcakes would sweat and be ruined, especially if it remained humid or if the weather turned bleak. By eight a sweltering sun had burnt off the heavy clouds which allowed the Southern Highlands initial show day to commence under a brilliantly clear azure sky.

Everyone in Yass praised the sun's warmth while decorating their horse-drawn floats with flowers on which youngsters would stand, decked out in their colourful costumes. Vintage cars in convoy would follow the floats to Cootamundra's showground.

28

Four days later on Saturday: Having won the first and second races with two thoroughbreds at the local gymkhana, Bjorn's stayers were scheduled to run on Canberra's elite racetrack at Lyneham. With the three horses strapped in their floats, he and his crew headed for the Australian capital at first light.

Leaving Gigi asleep, Kendall's car left Abergeldie at eight. He and Brianna collected Kirsten from Jalna on their way south.

On completion of the second race Brianna won a packet on Bjorn's horse Brazen Lass. Betting straight out on the tote proved exceedingly prosperous. On a winning streak and with the third race underway her funds looked like being boosted beyond all their expectations.

Kendall hadn't fared well, until Delta Boy came flying home in the fourth race. 'This stroke of luck has replenished my pocket. If Bjorn's other horses run true to form Bridy, I should be able to shout myself a *new* car. My old rattletrap is buried in the graveyard of damaged vehicles. And this second-hand car will be yours, while I'm overseas.'

Brianna smiled. Her purse was bursting at the seams. With another of Bjorn's horses pipping its opponents at the post, her future seemed assured.

The stallion Golden Promise, also trained by Buck Masters, had never won until today. At eighty-to-one straight out, it paid a good dividend on the tote. Bjorn was ecstatic. His trainer's hard work had proved beneficial to all pockets.

In the mounting yard Kirsten, dressed in her best finery, stood talking to Buck Masters while they waited for her husband and their jockey to join them on the podium. Delta Boy would receive the sash and rug for winning The Local Guineas. This was the stallion's second win within a month.

Sitting in the Members Stand, Brianna was trying to find her betting tickets. She'd won well over five hundred and sixty dollars. Smackaroos Bjorn called her winnings.

Kendall suffered a substantial loss in the third last race, until he forked out a fortune on the trifecta. The horses of an unknown origin he'd picked at random, came flying home. 'Lady Luck hasn't completely deserted me, Bridy. I don't have your knack of picking the right stayers though, darling.'

'I'm usually unlucky in most things. Although I've done well today, so has Kirsten. It looks as if we'll be going home with our pockets and purses bulging with money.'

'Bridy, I'm going to place four bets on this race. While I'm there I'll give my two friends another buzz. One of them should answer their home phone.'

Kendall didn't explain why to his fiancée, she knew who he meant. From then on, the horses he picked came home first and at a fiery pace. He couldn't believe how good luck could bolster his wallet in such a short time.

Capping their day, Brianna won "The Ladies Fashion Stakes". Smartly attired in a navy linen suit, her V-neck white blouse peeped from under her collarless hip-length jacket. An enormous spray of fine ostrich feathers, speckled with miniature spots of dove-grey, adorned her left shoulder. These plumes swept up to meet soft-burgundy feathers of her navy hat. A similar spray beneath its large brim curved down to meet the shoulder spray. With every movement these delicate plumes formed an S-shaped adornment. Her gloves, shoes and handbag were also crafted in dove-grey leather.

Kendal's eyes settled on his fiancée, while escorting her across the damp turf to the winner's stand. Brianna looked stunning. Pride boosted his morale as she elegantly stepped up onto the podium to accept her prize from the Mayor.

Nonetheless attractive, Kirsten's elegance and graceful walk displayed her silver-blue faille suit and apricot blouse to their advantage. Her hair adornment, composed of a huge chiffon rose in soft muted tones, similar colours of her suit and chiffon blouse highlighted her complexion. A silk veil cascaded down over her forehead. In the rising breeze this flimsy fabric fluttered slightly as her hips moved to the cadence of gentle music filtering through a series of loud speakers.

After collecting their individual winnings, Bjorn noticed Kendall looked subdued. He surmised his quietness related to what his father had called his mother, a bitch and a whore. In retrospect, Bjorn put this young physician's depressed state down to his mother's insistence of staying home on Abergeldie.

If the truth be known Kendall feared his father may be lingering somewhere in their local vicinity. He began to think while leafing through his winnings. *Just the idea of him being anywhere near my mother gives me an apprehensive feeling. I don't trust him and I never have, or will. May he rot in hell!*

Bjorn assumed Kendall was worried, knowing his father might take advantage of their absence. They both knew Gigi couldn't defend herself against his superior strength and savage blows. Should he return home while no one was there to protect her, Ian would grab the advantage to kill his wife. Thinking of this reminded Bjorn of the night when he and Jamie had found her close to death, hidden under branches in Jalna's top paddock, adjacent to Abergeldie's fence.

In between eating and drinking their cool, non-alcoholic beverages, Kendall recalled speaking to friends in Yass. The initial phone call was to Jarvis, then to Peter Bucknell whom he asked to drop in unannounced on Gigi. The diplomat's car pulled in Abergeldie's drive as Jarvis arrived. With plausible excuses, the men pretended they were paying her a courtesy call.

Gigi didn't have a clue that her son was the instigator of their visit. In a dismal mood and with her mind focused on a traumatic past, she welcomed these friends with the grace and elegance befitting her status as an important woman in their town of Yass.

Both these gentlemen plied her with flowers, chocolates and small presents. These sentimental gestures and endearments greatly boosted her mood. With their kind thoughts, care and encouragement, her melancholic noontide developed into a joyous afternoon.

At the conclusion of today's racing carnival, Bjorn's crew harnessed two of his horses in the dual float. With Delta Boy secured in the single horse float, their handlers departed Canberra and headed north.

The Svenssons and their friends went to an exclusive restaurant in Lyneham to celebrate. All were well endowed with their winnings of the day. As was Bjorn's custom, he shouted everyone to dinner, including Buck Masters, his horse trainer.

Their journey home proved wearisome and long. Caught in traffic, their trip didn't conclude until midnight. Nevertheless, everyone had enjoyed the day and their delicious meals and the service couldn't be faulted.

Kendall looked surprised to see a couple of vehicles parked near the house. 'We worried needlessly over Mum being alone all day. Bridy, I hope she hasn't suffered a relapse or a stroke. The first is Stephen's car and the other vehicle is Peter Bucknell's. His sly smile didn't fool Brianna.

'You knew they would both be here, Kendall. I twigged what mischief you were up to, as soon as you made those two phone calls.'

29

On Friday of that week Father Brady returned from his short visit to the Snowy Mountains. Inspired by the once majestic countryside surrounding the small town of Tumut, he found the valley air embracing. Interested in the different lifestyles of the townspeople and his hosts, he had enjoyed their gracious hospitality.

A peaceful ambience of the quiet valley had inspired him to wander ore the hills and local pastures. Hills sparsely speckled with green were burnt to crisp straw under a blazing sun. Swift flowing streams, now reduced to rivulets with water gently trickling over moss-covered rocks disappointed and appalled him. This Irish priest revelled in driving between serene and peaceful fields, carpeted with purple Patterson's Curse, a vibrant and destructive weed, and a killer of animals. He thrived on the hospitality of local folk who welcomed him into their homes. Their generosity bridged horizons and made him feel homesick.

With a laden heart, he waved farewell to his hosts whose home nestled amid red-ochre hills of the Tumut Valley. Travelling in the bus back to Yass, his heart yearned for Abergeldie and the family of whom he'd grown so fond. These once green grasslands, parched of nourishment, now looked barren. In no respect did they resemble his beloved Emerald Isle.

A welcoming committee greeted him on Abergeldie's front steps. Brianna accepted his suitcase while Kendall assisted him up onto the verandah. Its bull-nose iron roof usually protected them from the powerful sun rays.

'It's good to have you home again, Father George. I see you may've brought the rain with you. Those clouds rimmed with purplish-black are laden with hail. Before dusk we'll be walking in puddles and singing in the rain.'

Gigi greeted him in the lounge as a fresh aroma of cakes tantalised his taste buds. 'We've just finished baking your favourite date scones. Harriet's daughter Rose and Brianna cooked them especially for you Father George. My meager effort was to bake a batch of lemon meringue tarts. They look edible, but may not be to your liking.'

Kendall glowered at his mother. 'Your lemon meringue and jam tarts are far better than what are sold in the local shops. Mum, you know I never tell you a fib.'

Her frown of dissention warned him to be silent. 'Well, be that as it may, Kendall, don't just stand there, get Father a drink. He's probably thirsty after the long trip home.' Gigi paused to re-tie her apron. 'Perhaps you might like a glass of homemade Ginger Beer or cold lager from the fridge, Father.'

The priest's whimsical smile amused Kendall. Neither of those refreshments suited his Irish palate. Draping his coat over a lounge-chair, he put his hat on top of it. Nobody minded and he detested a fuss of any kind.

Brianna joined the throng and the priest embraced her. 'Me darlin girl, there's something here fa you ta read. The bus stopped at a library before we reached Tumut and a lady looked on a machine and found an item in a German paper. She copied it fa me ta show Kendall.'

In a kind gesture Brianna touched his hand. 'Father, do you mind if I read it tomorrow? We've all had a busy day and Mrs Ross is very tired. She mentioned an early night. We will all be turning in not long after dinner.'

Brianna accepted a cold drink from Harriet and thanked her. Collecting her things from the coffee table, she gave the priest's cheek, a polite kiss of respect.

'Not at all, me girl, I relished the meals in private homes in Tumut. And I found the trip home tiring. The scenery was spectacular, even though the dry paddocks needed a good drop of God's precious rain.'

Excusing herself, Brianna hastened to ask Gigi not to show her photo albums tonight. They weren't urgent and could wait.

The priest nodded as Brianna left the lounge room. His gaze then reverted to Kendall who stood not a shoe-length from the lounge bar.

'Kendall me boy, would you mind pouring an old man a drop of ya delectable nectre? Me mouth's a bit dry. I dislike the taste for tea. I prefer me favourite Irish poteen, the brew I drink in me local pub at home.'

'I've already poured you a glass. I know you prefer your whiskey straight, Father…'

Mrs Graham interrupted him to announce dinner was ready.

'Kendall, if your family and guests aren't seated within minutes their meals will spoil. Braised leeks on pepper steak won't stay hot for long.' Slaving over a hot stove, baking Christmas cakes and a chocolate cake for Kendall's birthday, Harriet disapproved of their dinners getting cold.

Everyone acknowledged her request and washed their hands. With this worry over she returned to the kitchen to cut the iced plum pudding. Bowls of whipped cream and jugs filled with hot brandy sauce and custard sat on the dining room table.

10 pm: After joying a delicious meal, Gigi went to her room. Her guests sat in the lounge room to sip their coffee. Nattering over Father Brady's trip to Tumut and the previous week's incidents, most of the guests chose to leave. Vacating the lounge room, they congregated on Abergeldie's front verandah, where Kendall apologised for his mother's rudeness. Saying their goodbyes, the Svenssons also decided to call it a night. Kirsten waved as their car joined a convoy of vehicles heading towards the homestead's front gates.

In an amorous mood, Kendall coerced his fiancée to walk up the drive. The pup, Amber ran on ahead to squat near the cedar tree. Under a quadrillion of stars, they discussed their individual overseas flights. With their time together lessening by the hour, their personal differences must be confronted. 'You can't blame yourself for his evil misgivings, Kendall. Something must've caused your father to leave home that night. He deceived everyone here.'

'Empowered to kill, not deceive. That Nazi murdered innocent victims and he tried to murder my mother twice. That is why I hate him. I will *never* refer to *him* as a relative of mine. He doesn't deserve that paternal title.'

30

6pm: The following night: Feeling anxious and at a loss for something to do, Gigi wondered why their guests hadn't arrived. Attired in an exquisite gown of pale green, she chose to keep well clear of the kitchen. Harriet and her daughter, Rose with help from the shearer's cook were accomplished in culinary skills. This eased her mind and eliminated the hassle of having to worry about tonight's menu.

Rose had finished the salmon patties, hors d'oeuvres and prawn cocktails. And with the main meal well underway, Harriet decorated the trifle and peach tarts, a specialty of the patisserie in Paris where Pierre frequented. The staff cook had prepared a variance of savoury dishes, fresh salads topped with herbs from the garden. Red current jellies scented the juice of hot geranium leaves and raw beetroot set in aspic decorated with dill fronds looked delectable.

Feeling fragile after a ride, Gigi relaxed with her son and Brianna in the lounge room. Within minutes Father Brady joined his hosts who were nibbling savouries. With a piece of paper in his hand, he noticed his missal was missing. Standing beside Kendall he glanced at the brochures in his hand that described his Parisian and Continental destinations.

'Kendall, will ya please read this item? A library lady printed it from a German paper on a screen. I think it will interest ya. It did me. And it might help ya friend and the police to catch him.' The priest seemed a bit despondent.

'Father, do you mean this came from a German newspaper? I'll do it now, before our guests arrive. You've whetted my curiosity as to the year of its printing.' Kendall read the item and gasped. 'This is indeed interesting. It will be a huge asset to the Federal Police. If you don't mind Father, I'll give this to Peter Bucknell.'

'I knew it would interest ya, me boy. Pass it on by all means. I have a copy in me missal. But I can't find it. I must ta left it in me room.'

'No you didn't, Father George. I picked it up and put it in your room minutes ago. I'll pour you a nip of your favourite whiskey and then we can talk this matter over with Senator Bucknell, who's talking to my mother.'

'If ya don't mind Kendall, will ya make me whiskey a bit more than a wee dram? I am thirsty and ya can keep that scrap of paper. I won't need it.'

Kendall gave the housekeeper an appreciative nod. Sipping his iced coffee spiked with a drop of brandy at the bar, he poured a double nip in a whiskey glass and handed to Father Brady. He then excused himself and left to speak with Peter Bucknell.

'Read this photostat printout and tell me what you think, Peter.'

'This was printed in nineteen forty-four. There's no mistaking who the editor meant. This is an obituary of von Breusch, a revered Reich captain and his funeral. The Feds will love getting their paws on this. That's if they haven't already seen or procured a copy of this Hamburg bulletin. Makes one wonder how the person who discovered this item knew of its existence. Probably an educated librarian traced it through her computer.'

'Father Brady went to a library on his way south. He says a woman printed it from an old newspaper on her microfiche. Well, I gather that's what he meant. He won't be rushed while narrating his journey through Austria in the war years. Father was quite interested in this editorial last week, or so it seems. Unfortunately, I can recollect only part of his long retinue. Brianna's birthday and our engagement party absorbed a huge spot on everyone's agenda that night.'

From the hall, Gigi glowered at her son. It warned Kendall not to discuss his father's sudden departure, or the argument that had ensued prior to them leaving for Jalna. The cause of their argument she didn't wish to be exposed. The incident, now forgotten, must remain in the past. Listening to her son now, conflicted with their peaceful evening and swathed her heart in sorrow.

Keener than a fresh gust of wind, in breezed the Svenssons. Kirsten looked alluring in her mid-length silk gown of delicate tangerine. This enchanting colour complimented her olive complexion and brown hair, done in a French roll and held in place by silver combs. Bjorn felt comfortable in casual slacks of fawn linen, with an open-necked cotton shirt, a gentler shade of his wife's frock.

'Sorry we're late for dinner, darling. An urgent phone call detained Bjorn.'

'It's okay. Dinner will be served soon, Kirsten.' She nodded to Kendall. 'Put their coats on my bed, thanks son. Our other guests are also late.'

'Don't fuss, Mum. They'll be here soon. There's no need to panic.'

The ladies carried their drinks and sat on the swing to chat over this week's events. An oppressively hot night, the front verandah seemed an appropriate spot. A slight zephyr stirred the air enough to keep everyone comfortable. The delicate aroma from Abergeldie's circular rose garden embraced their senses. Golden poplar and silver birch leaves rustled to Nature's tune along the drive. As twilight mellowed their glossy pelt reflected a brilliant glow of the full moon cresting a cloudless horizon.

'The breeze is turning cool. You're shivering, Gigi. I'll get your warm cardigan.'

'No Kirsten, I'm going indoors,' Gigi shuddered. 'Apprehensive feeling runs right through me whenever I think of Kendall defending my honour by striking his father. I suppose his mother died years ago in Scotland. He's lied to me for years. And I can understand why Kendall hates him. How can I empathise with him, after lying to me about his mother's illness?'

'Forget Ian. He's not worth worrying over and making yourself stressed. Bjorn and I are staying on Abergeldie until the Federal Police and their investigative team arrive in a week. I could be wrong, it may be less.'

'They say old habits die hard.' Gigi smiled. 'Manny and I were very close Kirsty, even though you and I were working in Paris under Pierre's tutorial guidance.'

'We enjoyed dancing in ballets, with Pierre as our choreographer. He seldom missed my mistakes, or a false step in our routines. We managed to eat well and create our own fun on those summer days together in Paris and Milan.'

Brianna embraced their enthusiasm with a gleeful smile. She understood and envied their enthusiasm as dancers, although in a different field of the arts.

'Darling, we're all anxious to see your photos tonight. Two of my albums are on the coffee table. I need you to give me your honest opinion of them, Brianna dear. There's a photo of you Kirsten, taken with Pierre in his new Mercedes. And I don't mean his daughter That snap he took just before you left Montmartre and went to London.'

'I remember that night well, Gigi. It was the final time we danced together in the Parisian Ballet. You'll meet Pierre Bouvier soon Brianna, I'm sure you will be enchanted by his charms. You'll be fascinated by his tales of the operettas he choreographed. He's a caring gentleman, typical of old school and ballet tutors in elite Parisienne cities.'

Kirsten thumped the air with her fist. 'I forgot to ring Pierre last night. Gigi, remind me to give him or Mercedes a bell. That's if they're still in Paris. This week they were going to Le Touquet on business.'

'I've been meaning to phone Pierre and his family for ages. Somehow with all our worries I forgot. I may give them a buzz in the morning. Not tonight though.'

Gigi looked at hall clock. *If I phone Montmartre now, it will disturb everyone. Pierre's a busy man, and I doubt if he, Raoul and Mercedes will appreciate being awoken by me, in the bleak dawn hours.*

Brianna turned to speak to Kirsten, until she heard vehicles pulling into the drive.

7pm. Doctor Jarvis climbed from his car, followed by their local choirmaster Ted Hardacre who also alighted from his vehicle. The diplomat, Peter Bucknell wished to leave early tonight due to early commitments in Canberra on Parliamentary business.

With everyone seated in the dining room, Gigi nodded to her cook. Harriet carried in trays of savouries and finger food. Rose followed her mother with hot beverages for the ladies. The men preferred alcoholic pre-dinner drinks to have with Kendall, who posed as their waiter. This allowed his mother time to spend with her guests. Relaxing with her feet on a stool, Gigi sipped a light aperitif while her female guests discussed ballet routines and theatre plays. She knew it would be five minutes before Harriet and Rose carried their meals to the dining table, set with crystal goblets on a white cloth. The calm serenity in the room encouraged her guests enjoy dining under soft candlelight.

Harriet put platters of braised vegetables and beef sautéed in wine on the table for each person to choose what they preferred to eat. These were followed by luscious desserts of meringue souffles, peach tarts and fresh strawberries. Everyone conversed with their neighbour while eating their desert. Bowls of delicate mint slices and fruit bonbons or Peach Melba were placed in front of every guest.

On completion of the meal Kendall ushered their guests into the lounge room. This elegant room with delicate green furnishings created an ambience for relaxing. The lounge sofas of white and lemon-cretonne with navy cushions invited everyone to be seated. Ceiling fans on low circulated the air. Yet not a drift of paper moved off the mahogany-polished side tables. And linen table napkins on the small traymobile were undisturbed.

Father Brady yawned then asked Kendal for a drink of his favourite brew. 'Me boy, could ya pour me a wee drop, please? Then I might toddle off ta me bed.'

'Father, you're a little unsteady on your feet. I think it unwise for you to have another whiskey. There's a jug of iced water here and I can add a drop of lemon juice. Or I can pour you a glass of non-alcoholic wine. It tastes better than our local apple cider.'

'Cider with ice? No thank ya, Kendall.'

Waiting for Brianna to return from the lady's room, Gigi sipped an eggnog spiced with cinnamon. And with a clear view of the note Kendall was reading she was curious.

'What's so amusing, son. The look on your face says it's more horrible news.'

'Don't be a stickybeak, Mum. This letter has nothing to do with you. This is an advertisement for a doctor in Queenbeyan, a suburb of Canberra.'

Kendall hated lying to his mother. Silently he continued to read the item in his hand. *This verifies what we already know. It must've been stapled to the microfiche printout Father George showed me before dinner. This fell on his floor when I put his missal on the bed.*

If I read its faint transcribed text it might make sense. 'This citation and medal are awarded posthumously to Rudolph von Breusch for services rendered to our glorious Fuhrer, Adolph Hitler at the Reich Chancellery until 1938. The captain previously received the Iron Cross medal from Keizer Wilhelm at the end of the 1914-1918 war'. Kendall gasped in disbelief. 'This verifies that my grandfather worked at the original Berlin

Chancellery before it was destroyed by fire. Tomorrow, I'll buzz Peter in Canberra.'

Caught unawares, his mother interrupted him. 'Kendall, I'm speaking to you. You were deceitful for telling me the real reason for your father's departure two weeks ago.

'Mum, I wasn't deceitful. You know what occurred that night.' He kissed her cheek. 'I'm going to give Father George his medication, then I'll take the pup for a walk. After that I will turn in. Don't stay up late or you'll collapse. You're so damn tired you can hardly stand on those wobbly feet.'

Kendall wasn't quick enough to catch his mother who collapsed on the floor beside him. Kneeling, he could feel a threaded and wavering pulse then heartbeats strengthened.

Doctor Jarvis rushed to fetch his medical kit from his car. For some minutes they worked on their patient until Jarvis could feel a strong even pulse in her temple. He monitored her heartbeats until they no longer fluctuated and her rhythmical pulse steadied.

Kendall held a cool compress on her forehead and refused to leave until Stephen had finished assessing her vital signs.

'My God, her forehead is burning hot. Mum won't relapse into a coma, I hope. Stephen, I'm worried about her shallow breathing.'

'It's quite what I expected, because of the horrendous shocks she's had recently.' While talking Jarvis drew up a drug in a sterile syringe. 'Kendall, I'd be a liar if I denied being worried. With constant care and this drug I'll administer to your mother it will ease the palpitations. Settled in her own bed she'll be sleep for hours. Ancient apothecaries believed sleep was the best remedy to cure all ailments. And we both know that's true.'

Grave-faced, Kendall observed the colour in his mother's face return to normal. He expelled sighs of relief as her eyelids fluttered. Her pale cheeks, sunken with pain gradually returned to normal.

Stressed, Kendall left Jarvis to monitor his mother's condition. Feeling despondent, he returned to the lounge room to speak with their two remaining guests.

'Brianna, I'm going to brew a pot of hot tea. Would you care for a cup? We can drink it on the front verandah swing. Mum's just gone off and Jarvis is going to sleep in her room tonight. Stephen has advised me to go to bed. I might have a shower, before I turn in.'

She glanced down at a folded paper near her shoe and picked it up. 'Is this note yours, Kendall? It's not mine or your mothers. It could belong to one of the guests who've gone home?'

'Pass me the note please, darling? I'll give it to Peter Bucknell. Nothing important. I might ring him in the morning. Now two cups of tea coming up. There's a jar of Harriet's shortbread biscuits in the kitchen. She won't mind if we eat a couple.'

They ate the biscuits, Kendall's favourites, and drank their tea while cuddling on the front swing and neither commented on the note in his pocket. Brianna thought it strange why he wanted to phone the Canberra diplomat so soon after him leaving Abergeldie.

Oh it really isn't my business. Kendall will tell me if it's urgent. Brianna knew she should have told Kendall about his father's unwarranted and seductive look she received on her arrival.

'How could I say a thing to either him or his mother?' she uttered in the toilet. 'I told Father Brady and he agreed with me. The problem no longer exists, and they have suffered enough, without me adding to their worries. Now, I prefer to forget the incident.'

The shock of Gigi having seen her husband in a Nazi uniform, standing behind Hitler confused Brianna. *Kendall should have told me that was why she fainted.*

'I'm going to bed, darling. Don't wake me in the morning please. I'm overtired and I need to catch up on some sleep.'

'What have I done or said to upset you, Brianna? You're not usually crabby. All right I'm off to bed now. I'll see you in the morning after I've given Father his medications. Probably at breakfast.'

Kendall recalled how astounded his mother had looked on seeing his blood-splattered shirt. The image of his father in a Nazi uniform wearing a swastika armband was imprinted on his mind. It now took him a while to assimilate the actual meaning of Father Brady's latest microfiche snippet. 'If I can't contact Peter Bucknell by phone, I will post this note in my pocket and register it to his office. It'll probably take two days to reach Canberra. Blast I'm fed up with all this worry. I'm going to have a shower and then bed.'

Constantly thinking of these microfiches in bed and unable to sleep, Kendall uttered while re-reading the recent one, 'Bucknell told me every scrap of date you uncover is important to bring this rogue to trial. He's right. I'd be foolish not to copy and collate these pages as I get them. It will be a huge benefit to the Federal Police officers when they do arrive. I must ask Peter if he knows the exact day they will arrive. This microfiche data could be extremely important evidence and his colleagues will need them to confirm any sightings of the Nazi absconder. Why the hell did Father Brady look into this matter when he hated the Nazis? My mother only married that miserable sod to protect my future. Brianna mistrusted him from the first time that she almost bumped into him in the hall. Instead, he turned and headed to his den. I remember him abusing Mum over some triviality. His anger lies deep in the past. Wherever he is now, he cannot hurt or abuse her. I hope he rots in hell, because that's where he belongs. The Fed's in Canberra should have some trace of his whereabouts by this. Once I've spoken to Bucknell, I'll know something positive on that score.'

8am, next morning. Bjorn rang and spoke to Kendall on his way to breakfast 'Hi matie, we been a bit concerned over your mum's collapse. How is Gigi feeling this bright morning? Kirsty's on her way over with a surprise for her. Keep it under wraps or you'll spoil it. I'll pop in on my way to town. There's another few truckloads of hay and lucerne on the road from Wagga and one from Goulburn. Your sheep and cows and mine will have a good feed, if there's no holdups on those busy highways. See you around tennish mate, Better get going or I'll be in strife with the wife.'

'Mums the word, BJ. I'm just going in to see if she's awake. Things are moving this end. Tell you my news then. My breaky is just about ready, Bye.'

Neither he nor Kendall realised that Bucknell's secretary had photographed everything he'd received and handed it over to the Federal boys in Canberra. They went ballistic before logging the miniatures and photostatted letters in their computers. The AFP agents were building a solid dossier on the absconded Nazi, they now knew under the name of Rolf von Breusch. Their overseas researchers had an enormous amount of documentation and it was growing at a phenomenal rate daily. Time limited this ex-German officer's life who, during the war, was responsible for the deaths of more than a thousand persecuted Jewish families and over two hundred innocent victims of the catastrophic events that had occurred in Austria and Poland.

Neither Bjorn nor the two gentlemen in Abergeldie's lounge room had mentioned this to Kendall. They were more concerned over his mother's sudden collapse. Jarvis confirmed Gigi was conscious and functioning as well as could be expected.

'Struth, I now know what Kendall meant.' As his eyes focused on the copied miniatures resting in his palm, Bjorn felt ill. 'The likeness of Ian in that Nazi uniform during the war years is identical to him now. This is incredible. It's difficult for me and everyone to believe he actually was a Nazi.' After giving it more consideration, Bjorn smiled. *Yes, I can. I often heard him say an odd word in German when he's been four sheets to the wind. I was right. Now I'm all ears to discover as to where that bastard has flown. I bet Gigi didn't have a clue about his Nazi past. Well, not until she got a glimpse the damn photos two weeks ago.*

Looking as if he'd seen a ghost, Father Brady appeared in the doorway. His features lacked colour and his grave expression said it far better than words could. From what he'd overheard, he suspected his miniatures had reincarnated the past and reinstated it in this room.

'Tell me, why the fuss over them snaps?' The priest was confused. 'They're only photos of the Nazi officer I showed ya all, before I went to the retreat in Goulburn. Kapitan von Breusch, like all Reich officers, often had their photographs taken with Hitler fa propaganda purposes. I shoulda destroyed those years gone now.'

'Father, sit down until Kendall comes back. I'll let him explain everything to you. Give me your glass and I'll pour you a fresh whiskey.' Bjorn filled the goblet close to its brim and passing it back he warned, 'Sip it slowly or you might choke. I've drained the bottle and I doubt if Kendall has another one in stock. There's a bottle of cheap Scotch on this shelf. But then it wouldn't have the kick of your Irish brew.' *What a liar you can be at times Svensson,* he mused unconsciously fingering the miniatures now tucked in his shirt pocket.

A clandestine glance at the Irishman's pallid-face made Bjorn cringe. Rampageous ideas racing through his mind fired him into thinking: *This poor bastard has suffered enough without arousing the dragon within for a second time. The photos in my pocket and that letter, engraved with the*

Nazi emblem now in his missal, have existed all these years and are still in a pristine condition, which is impossible to believe.

Fixing himself a stiff brandy Bjorn mentally exploded. *The German whom Kendall and I prefer not to name was a born mongrel. Our government will undoubtingly deport that king of all bastards when the conniving liar is captured. They should hang, not deport him. Kirsten and I are to blame for sponsoring him and Gigi to live in Australia. Why she chose a Nazi to marry makes me want to puke. I thought he had Germanic heritage, because of his dictatorial attitude. Now the miniatures have verified my suspicions.*

Utterly bewildered, Father Brady looked at his host, who on re-entering the lounge room sat down beside him.

'Kendall me boy, how's ya mother feelin? Quite poorly I suppose. I never meant ta frighten the dear lady with those stupid photos. I didn't mean ta upset her.' The priest pointed to a faint blemish above his left cheek. 'Look at the scar. Well, that cruel Nazi officer struck me there with his wooden baton.' He kept repeating this phrase in sincere and soft monotones.

Kneeling down, Kendall placed his hand on his knee and spoke in a gentle, though concerned voice. 'It's all right Father, we know what he did. Please forget him for now, or you'll make yourself ill. We understand your anxiety and it's natural under the circumstances. Finish your drink and then I'll walk with you to your room. It's time to administer your injection of Diazepam. Then I'll let you rest until morning tea is ready. You asked about my mother. She is asleep and Doctor Jarvis is going to remain in her room until I relieve him in an hour.'

On his return from settling the priest down for a rest, Kendall plonked his bottom on the lounge sofa to drink his now cold cocoa. He did want coffee this morning.

'My God, this latest news will most likely kill your mother. No wonder she collapsed again. I'll cut his throat, if I ever catch that Nazi absconder. This is not a treat, matie. It is my unsaid promise to your mum.'

'Now both the women and Father George are resting, I'll tell you latest news, Bjorn.'

'Why! What are you talking about? We've nutted over all this nonsense before…'

Kendall cut him short. 'Peter Bucknell, with his official expertise, will know the right procedures to follow with theses scraps of paper. Father George probed into some history in or near Canberra on his recent trip south. A librarian discovered this data on her computer and printed out a copy for him. He has no idea of the repercussions this will have on tracing our absconder. I didn't know Farther had these pages, before Peter left for Canberra last night. Try not to put grease on that signature Bjorn.'

'What are you crapping on about Kendall? I washed my hands, they're clean. I only ate a slice of toast that Harriet made me topped with her pineapple honey on it. And that was yonks ago.

A shaved, grinning face appeared from the main hall. 'May I intrude on your privacy, gentlemen? You did pass everything to me. Kendall. And I have the copies for you here. At present the original photographs and letters are a vital part of a huge dossier the Feds are building to track down your absconder.'

'What a relief.' Bjorn looked at Kendall. 'We quite expected the Feds to put a trace on him. It appears to me they could be looking in the wrong direction. This small chit says he may be lurking in or around Kings Cross in Darlinghurst, an inner suburb of Sydney. Not Chinatown as you suspected, Kendall.'

'Bjorn, how did you come by that film pass?'

'I went snooping last night and found it the den. In your search you missed seeing this torn stub under the desk. Trust me ta find it. I'll let our friend here have a brief gander at it, before he darts off to somewhere else.'

The indignant frowns of Bucknell and Kendall warned Bjorn not to continue. 'What we don't need now is someone sprouting rubbish, of which he knows naught. Sorry if I sounded abrupt, Bjorn,' confessed Peter Bucknell. 'Every bit of vital evidence will be beneficial to my Canberra colleagues and the Federal Police. Even ASIO have a huge stake in this project. Their agents have a colossal dossier on von Breusch, which I've actually seen in recent days. Everything we've discussed in this room must not be repeated. If one word is, there will be a heavy penalty that person will face. Security is the main word in this conversation.' Peter knew Bjorn wouldn't dare say a thing. Nevertheless, he didn't regret quoting this little excerpt from the Federal official manual.

To soften the blow, Kendall produced a couple of folded items from his wallet. 'These will capture your interest, Peter. You saved me the effort of posting them to your official office. They will clarify a lot of what we suspected. Microfiche copies of official Reich documents. I never suspected that my grandfather also worked with our absconder. Well, my wretched father. There, I've finally acknowledged *him*. Don't ask, or expect me to say it again. I won't.'

'These letters may also interest you, Peter. I *am* guilty of not handing them to our local bloke. Kirsten keeps nagging at me to drop them into the Post Office. Busy, I keep forgetting. You'll find some safe spot to put those letters until you fly back to Canberra. I haven't opened them, as you can see, if you turn them over.'

'The name M Kleinhardt doesn't ring a bell with me. And as you all know I've lived in Yass for well over two decades. A strange surname. It sounds German. I don't know anyone around here with this name. There's a Dieter Smidt who lives further along my road. These letters in my hand could belong to someone he knows in town. The Federal boys will have hours of fun researching this lot.' Pulling a folder from his overcoat pocket, he handed it to Kendall. 'Open that before you comment.'

A paperknife taken off the lounge table was put into action. 'If I'm not mistaken, this folder contains two first-class tickets to Europe. I haven't cancelled my flight to Paris, neither has Brianna to Ireland. What the duce is going on here, Peter.'

'At your mother's request, I reorganised both your flights, Kendall. I've also arranged your new accommodation. I believe your group of friends left Sydney last Sunday for their six-week holiday, skiing in the Swiss Alps.'

Confused he stammered, 'Yes…yes they did, Peter. I intended to cancel my flight and holiday plans tomorrow. It seems you've done it for me. What about Brianna's flight home to Ireland?

'Your mother and fiancée both agree with me. Brianna asked me to rebook her flight to London with your ex-company, Bjorn. Everything's rescheduled and correct, Kendall. Ask the ladies if you doubt my competence and sincerity.'

'It seems you and Stephen have done your best to accommodate my mother's whims. Although, I admit she deserves your attention. Mum has suffered unnecessarily for months. Now she's free to live her life here in peace.'

'Don't be too hasty. There are a lot of contentious issues in the wind which may give you reason to rethink that opinion. I'm prohibited from divulging Federal policies to anyone here and in Canberra. The moment I hear anything positive, you will be the first to know, Kendall. Oh, about the top paddock. The Cessna, loaded with Federal personnel and technical equipment, is due to land at six in the morning. I'll need transport to drive up there, with you acting as my personal aide.'

'Am I to gather this is a hush-hush situation? I trust every member of our staff to keep their mouths shut, Peter. The same goes for my crew and Bjorn's shearers working in our woolsheds. Not forgetting the local blokes running our mob of sheep in the dry paddocks. Bjorn's blokes are trustworthy. Their livelihoods are modelled on our ideal principles.'

Bucknell nodded in agreement. The owner of a grazing property, he understood their predicament well. 'I concur with you both on this topic. It isn't all beer and skittles working under a harsh summer sun and in all weathers. The colonel will lecture every member of your staff about the rules to follow, while his team is stationed here on Abergeldie. I advise you and your medical colleague to come up with a foolproof plan to keep strangers, or nosy parkers at bay, Kendall. Let me know your decision as soon as possible. Then there will be no room for confusion, on either of your properties.'

Kendall laughed. *I won't have to worry. If all goes well, I'll be overseas for five or six weeks. I won't miss my mother chiding me. But I will miss Brianna's delightful company. Still, one must make the best of difficult situations and face the truth. Stephen will be here to administer Father George's and Mum's medication. I have something positive to look forward to on my arrival home. A secure and well-paid job in his practice in Yass. Oh, I must speak to Peter about in-house security, before I leave for Europe. I'll leave a contact number with Pierre Bouvier in Montmartre, in Paris in case he needs me.*

In transit to the priest's room Kendall spoke to his fiancée outside hers. 'I have Father's medications ready, I won't be long, Brianna. His health has improved since he returned from the seminar and that short break in Tumut. He probably found it a reprieve from the droll life he's seen here on Abergeldie. Father George has no idea that you and Mum arranged those trips though our local priest.'

Brianna looked at Kendall with tears in her eyes 'Darling, will Doctor Jarvis be staying here to monitor your mother's health, while we are overseas? I dread leaving without some assurance, in case she suffers a relapse. Your mum insists on me calling her Gigi, and I feel it's discourteous.'

'Let's discuss this together on the front swing. I'll be there as soon as I change the dressing on Father's arm. A slight abrasion, underlying a bruised and inflamed tendon.'

'Father says the fall occurred on his trip home from the Snowy Mountains. Kendall, I'm a bit skeptical whether that is the truth. All my life he's drummed into me never to tell a lie. Now, I think he's told us a couple of huge fibs.'

In his own bedroom, Kendall and Bjorn Svensson discussed the outcome of a phone call Senator Bucknell had made to his secretary in Canberra. 'We'll have to sweat on Miss Sanderson contacting him tomorrow. Peter said Dee sounded agitated when he spoke to her. Kendall, what can we do? It's not our problem. I think Peter blames himself for not advising you before he rescheduled both your overseas flights. I could've rearranged everything through my ex-bosses in British Airways.'

'Thanks for nothing. No, I shouldn't be critical of your motives, Bjorn. If we stand here chatting we *will* disturb Mum.' Kendall bid his neighbour goodnight and nonchalantly shook his head. 'I'll see you first thing. And thanks for being a good listener. I needed to confirm a few things, before Brianna and I leave here at 6 am this coming Sunday.'

'Kendall, I'll creep into the bedroom and give my wife a kiss. I knew Kirsty would be staying with your mother overnight. I'll be over early and we can put this problem to rest. By then Bucknell should have loads of information on the Cessna's E.S.T tomorrow morning. I've warned my crew to keep their mouths buckled on what they think may have occurred here. More importantly, of what the future might bring. There will definitely be no leaks from Jalna. I can assure you of that, matie.'

These arrangements suited Kendall. *From what Peter's secretary has inferred, he should be here around breakfast. Neither he nor Dee are familiar with my agenda for tomorrow.*

Kendall needed time to rethink the details of events thus far and how to confront another unexpected problem. Befuddled brain-wise, he tried to make sense of the upheaval that had engulfed their family for longer than three weeks. In turmoil, his brain refused to function in a logical capacity. Everything seemed to be working against this dedicated physician. Under stress his throat seized and he found it difficult to breathe or swallow.

Doctor Jarvis walked down the hall after taking Gigi's blood pressure and pulse. Having foreseen this dilemma, he tried to console Kendall by offering advice.

'Let Peter resolve this without your or my interference. He deals with difficulties of this calibre on a daily basis and knows what protocol demands. Besides, he's aware of correct procedures to follow to get results. Kendall, I know how difficult this must be for you to understand your father's unscrupulous life in Germany. Now you must forget both it and him if you want to retain your sanity. Your mother will need you to be here for her once Christmas is over. There'll be long periods of uncertainty to face in the coming year.'

Jarvis walked to the bar, and poured Kendall a whiskey sour. 'Here, drink this. A nip won't hurt you and it'll catch your breath for a second. You'll breathe easier then. Kendall, with all the stress you've endured and finding out the truth has shocked you. Having known you from your teens, I'm convinced that neither you nor Gigi will let him drag you down through a quagmire of misery. Believe me, he's not worth it.'

One gulp and not a drop of whiskey remained in the tumbler. Until the burning sensation eased, Kendall could neither breathe, nor speak.

'Hell, I'll never do that again. Stephen, you almost killed me. Struth, my eyes are watering. You damn well knew that would happen.'

'You're a wimp, Doctor. If nothing more that drop will make you sleep. Will it be okay if I rang a friend of mine, before he goes to bed?'

'Yes, of course. You can use the phone in Mum's office anytime. It's quiet and neither she nor Brianna will hear you from there.'

Before Stephen Jarvis headed for his bedroom, they enjoyed a spirited joke together. The topic of his father's Germanic life and his betrayal that Kendall dreaded never resurfaced.

31

Tempted to destroy the photocopies, he knew it would be foolish. It could hinder any chance of the Feds exerting pressure on their men to find and capture his father. Either way, Kendall vowed that his mother would never lay eyes on the ghastly photos again. He decided to keep them in the wall safe in his father's office, or den. No officer or stranger would think of looking for the incriminating photos and the microfiche copies in there.

Kendall realised his mother might not recover after this collapse knowing the man whom she married was an escaped war criminal. This was more than conceivable because her nightmares lingered on awakening. Frightening fantasies so deadly that they persisted for hours during the day and he feared for her life.

Ready to turn in and seeing Father Brady's light under his door. Kendall softly knuckled the wood.

'Come in me boy. I'm just readin me missal, Kendall.'

'I won't disturb you for long. I just popped in to see how you are. That ugly business tonight has upset you greatly, Father. I am sorry that you were caught in our misery. Good intentions gone awry.'

'Heaven forgive me, what have I done? I never meant ta create havoc or harm ya mum, Kendall. I shoulda burnt them photos long ago. How can I repay the damage I've caused?' the priest wailed, his eyes overflowing. 'Will she ever forgive me fa bringin the horrible miniatures inta your home?'

'Shush Father, the problem doesn't exist. It's over. Please try to rest.'

Kendall reproached himself for tackling the subject they both detested. To hear this man apologising for something that wasn't entirely his fault was gut-wrenching.

'Father, there's nothing to forgive. You did the right thing by disclosing your traumatic and turbulent life in Germany. It rebounded, but not because of your good intentions. Things have a way of affecting innocent people who least deserve to be hurt. You, my mother and me are all victims of that man's evilness. Forget this nonsense of taking the blame. It's more my fault more than yours. I should have known how revisiting the tortuous hours you spent in the Chancellery cell would affect you.'

Kendall refrained from speaking due to the restriction of his throat. Able to swallow he continued, 'Father, you weren't to know how that Nazi hated everyone he came across. He glorified in murder and killing innocent people.'

Having expressed his opinion honestly, Kendall reflected on his last statement. *It would terrible if this devoted and caring man disliked me, or suffered a nervous collapse because of my stupidity. I understand how our damn situation has affected him, more than he will admit.*

A long silence elapsed before he could speak to the priest in a comforting tone. 'Tomorrow you'll look at everything with a positive attitude. Guided by your encouragement and your prayers, my mother will be able to cope with this shocking dilemma. That should boost your morale. Keep in mind, I will be here until Sunday if you need to talk with someone who understands your anguish. Never reproach yourself for not destroying the miniatures.'

The priest began to doze and Kendall crept from his room. In a receptive mood he checked on his mother. In a mellow tone he sighed, 'The drug Stephen has given her is working, thank God.'

'Now I'll look in on Brianna. I know she was suffering from the lack of sleep. She's on the verge of drifting off so I won't disturb her.' Kendall edged her door to a close.

Kirsten dozed in the chair by Gigi's bed. Around four she awoke and noticed that her breathing seemed irregular. Fearing she might have a stroke, she disturbed Kendall. From four-thirty on neither of them moved an inch from his mother's bedside.

6 am. As the hall clock chimed the hour, it and the phone awakened Kendall. Struggling to his feet he pushed the bedroom chair and stretched, before attempting to answer the persistent buzzing.

'Doctor Ross speaking. How can I be of assistance to you, Madam?'

'I'm sorry to disturb you at such an ungodly hour, Doctor. I'm a junior partner in Stephen's surgery and I've phoned to say, myself and one of our locums have agreed to manage his practice for several weeks. Pass my message to him in you don't mind, please. Then he won't have to worry. This will allow him to spend more time with your mother.'

Suddenly it occurred to Kendall the reason for this early call from a stranger. 'Now I know why Stephen asked to use the phone last night. I'll string him along, before passing on this message and spoil his breakfast for a change.'

The moment their meal had concluded Brianna awoke screaming. Her nightmare seemed intensely vivid. In her dream she saw an officer in a Reich uniform standing over Kendall with a rifle aimed at his chest. He then bent over his inert body on the road.

Fearing she would awaken his mother, he and Jarvis hurried to Brianna's room.

Once they stabilised his fiancée, Kendall took his colleague's advice and returned to the kitchen, looked at the remnants of his birthday cake and sighed. Pensive, he knew Brianna would be dozing after the valium Jarvis had injected in her bottom.

To keep his mind occupied, Kendall carried Father George in his breakfast. As he approached the front hall a strange noise caused him pause and listen. A muffled sob came from the priest's room.

On investigating, he found him in a virtual mess. Incapable of absorbing a word, he was wallowing in self-pity. Kendall felt useless to help this man who constantly rocked while cradling his knees. The priest had become a victim of his own misfortune.

'I caused ya mother's collapse. Can't ya see me boy, it's all me fault? I never meant ta bring shame on ya family. Ooh laddie, I canna tell ya how sorry I am.'

Kendall understood his remorse. Holding the full glass of orange juice to his lips, it couldn't spill. Within several seconds Father George held it in his trembling hands. One good swig, he spluttered and almost choked.

'Why did ya give a man that stuff?' he wailed. 'Ya know I dislike it, Kendall. Why did ya not put a drop of Usquebaugh, or poteen in this for an elderly clergy?' Father put the glass back on the tray. 'What's a drink without a good drop of ya delectable nectar? Irish whiskey tastes better than juice.'

'You should not mix liquor with medications. Father, drugs and alcohol make a lethal cocktail and taken together they produce undesirable consequences.'

His hands flanked the air. 'Kendall, no more please? That stuff is vile.'

'Father, try to eat the omelet. It has ham, onions and your favourite cheese, topped with finely-chopped chives. You don't want to offend Rose, do you? She made this especially for you. And this hot toast. Eat your breakfast and then rest. Senator Bucknell is due to arrive soon. I'll tell him you're not to be disturbed.' A gentle hand rested on the priest's shoulder. 'Expel all those ugly thoughts from your mind and have courage in your faith. You shared your prayers with us when we needed comforting.' Without another word Kendall closed the door.

In peaceful repose on the front verandah, closeted thoughts surfaced. *What will happen when and if that Nazi absconder is caught? Never will I allow my mother to testify in court. The anguish of such an experience would be enough to drive Mum, and every sane person here mental. Losing her sanity because of his infamous past is too horrific to contemplate. Brianna has grown quite fond of my mother. It will be a huge relief off my mind while I'm travelling through France and Europe.*

He gave a long sigh then uttered to the pup lying at his feet. 'If a trial were to eventuate I will be unable to help my Mother. What ill we humans perpetrate against the innocent of this cruel world, to sanctify our own greed and lustful dreams. Come on Amber, we can't be lazy.' The pup jumped up, wagged her tail and looked at the lead in his hand.

The idea of strangers intruding on our peace is enough to make me cry. What a waste of time and energy that would be. He took a deep long breath. *Silence in the days of hell we must still endure will be our sanctity. Hate is futile.*

Basking in the glory of solitude, Kendall gazed up at a wide abyss of blue. Slowly his eyes lowered to red-soaked earth channeling mud down from the homestead's gates. Meditating, he smiled. *This rain is a godsend, a blessing to drown all my miseries, along with the sad memories of yesteryear.*

A flock of pink and grey galas drifted on the wing as he voiced in a soft, dismal tone, 'Peace will never return to Abergeldie. Nor to my mother or Brianna, both of whom I adore. Nothing will ever be the same again. How can our lives evolve to promising futures? All those I love and respect must suffer the indiscretions of a man whom I detest. Now the innocent must pay for his evil deeds.'

His eyes focused on Jalna's home paddock as the utterances ceased. *Deep in my heart, I will miss Abergeldie and the Svenssons, whom Mum and I value as our sincerest friends. They taught me how to live an honest and productive life and to respect my elders. Now they too must suffer. My family and our staff will be the recipients of a cruel defiler, a Nazi who does not know how to value life, or the true meaning of love. Where will this journey lead us, where will it all end?*

32

Anxious to see how his mother had fared in his absence, Kendall, with Brianna close on his heels, hurried up Abergeldie's steps. Kirsten and the men followed. They were all weary, but their contented faces greeted Gigi at the front door.

'Mum, you shouldn't be up this late. I thought you'd be asleep long before now.'

'Well, that's a nice greeting. Can't you do better than that, Kendall? How about a kiss and stop chiding your mother?' She laughed at the stunned look on his face. 'Come on in everybody. You all know my guests.' Gigi's smile turned to a scowl. 'Oh and you won't know anything about their sudden arrival either, would you Kendall?'

'Me! No. Why pick on me, Mother?' came his abrupt reply.

'Enough son,' Gigi snapped 'You gave yourself away by calling me Mother. It's a trait you've shown since infancy, when you were annoyed with me. Now, if you don't mind Kendall, please attend to our guests. I'm off duty.'

Gigi complemented Kirsten and Brianna on their choice of clothes and chatted over today's outing. They let the men discuss more serious matters on their minds.

'I'm glad you all had a prosperous day. Winning two thousand dollars on Bjorn's horses, plus his others sounds a good day's pay to me. What did my son win, Brianna? If he's like me, he's probably come home with empty pockets.'

Kendall, who heard from the lounge room bar, sniggered.

'Not quite Mum, I made a big hit on Delta Boy. Brianna put a trifecta on the fifth race for me. All the horses she picked came flying home. Mother dear, my wallet is bulging and is hard to close. Full with crisp hundred dollar notes.'

'Gigi, feast your eyes on the girl who took out The Fashion Stakes.' Bjorn nodded to Brianna who smiled graciously. 'And I don't mean no nag. Brianna won first prize and a huge box of exclusive chocolates. We're all so proud of her.'

'Bjorn, I'm not surprised. Brianna always dresses elegantly. I wish I'd felt up to going with you to Canberra. I didn't. Never mind, I fared the best of you lot. These very courteous gentlemen have kowtowed to my every wish.' Gigi winked at her guests. 'Peter and Stephen offered to keep me company until you lot came home. We shared a few jokes about the party. They were clean jokes, Kendall. Not like yours are at times. Then we drifted on to an uglier subject, one which I prefer not be reminded of tonight.'

Aware to whom Gigi was referring, Kendall thought, *I hope he drops dead before Christmas.* Disgusted, he switched topics and enquired of their guests, 'Gentlemen, would you care for wine or a nip of my prized whiskey?'

Peter Bucknell chimed in, 'Coffee for me, thanks Kendall. I've ingested enough liquor tonight and I'm driving home.'

'Same for me,' Jarvis piped up. 'Make mine black and one.' With an afterthought Stephen added, 'Kendall, I want to speak with you regarding my surgery. Make sure I don't forget, please. It will be to your advantage.'

'Okay Steven, I'll fetch the hot drinks. Going by that last burst of thunder, it looks as if more rain is on the way over those escarpments to the south-west. Be good if it does rain. Although we don't need a downpour to carve deep channels in the topsoil. It will create huge chasms and cause strife for the sheep. If they wallow in mud, their thick coats weighed down with water will cripple them. The relieving shearers will be here tomorrow at six.'

'Being truthful, I dread the rain. The main judging will be tomorrow. And you all know what hell it is leading cattle and animals around the ring in wet weather.' Bjorn put in his penny worth quite bluntly to say how he could recall Kendall's horse throwing him in last year's equestrian event. 'Gee, you were a sodden mess.'

Seeing a funny side of that predicament Kendall laughed. It livened the evening, or what was left of it. 'Where has the night gone?' he grimaced.

Thanking Gigi for an enjoyable evening, Stephen Jarvis needed to confront her son regarding a locum.

'Spare me a minute by the car, Kendall? Good night my gracious hostess. I must be making tracks, I have a full surgery in the morning. And operations tomorrow afternoon. I'll be out on my feet, if I don't soon leave.'

'Goodnight Stephen. And thanks for making it an enjoyable evening for Gigi. It won't be long before I hit the sack.' Bjorn yawned. 'Tomorrow's going to be a corker for me. Grooming horses for the local show take my boys hours. The rams must be perfectly combed and brushed before parading them in the ring. This year I want to double my display of first prize ribbons.'

'I'm not staying either, Gigi. I've enjoyed this evening with you and your charming physician.' The diplomat, Peter Bucknell kissed her cheek. 'I'll drop by tomorrow before I leave for Canberra. I'll be gone a week. And I promise to bring you more chocolates from our city courtier. Pity I can't bring you the ones I buy from an emporium in London. Theirs are absolutely delicious. Customs, you understand.'

Gigi thanked Senator Bucknell for the box he bought her this evening. With a laden heart she watched him leave. *He's a gentleman and his manners are impeccable. He considers me a lady of substance and knows how one should be treated. Not like that barbaric husband of mine, whose irascible manner I found intolerable. Oh damn, why did I have to think of him on such a marvelous evening with our friends?*

Feeling jaded, Gigi spoke to Father Brady for a moment. Excusing herself, she left Stephen Jarvis to talk over medical topics with her son.

Kendall appraised the rising mists while leaning on their front verandah railing. 'You wanted to discuss something with me, Stephen. What's on your mind?'

'You told me some time ago you intended to work up north in a new practice. I will have a vacancy in my surgery soon. Think it over Kendall, we work well together, you and I. You'll have your own fully equipped surgery to attend your patients. It's air-conditioned and quiet with a pleasant view overlooking the river.'

'It's odd you should mention it now, Stephen. I received a letter today stating the position will still be mine, once their surgery is rebuilt. The physicians in Townsville are working under hardship in a small building at present.'

Kendall considered his proposal for a second, then agreed to give Stephen's offer some thought while on his overseas holiday. 'I hate being a paper jockey sitting on my bum here. I'm not cut out to be a land-donkey either. Nor a filing clerk like Mum expects me to be twelve hours a day.'

'You have two weeks to consider my offer. Then I must have something concrete before I advertise for a young locum.'

Stephen Jarvis knew Kendall's future plans included specialising in pediatrics or obstetrics. *Studying infants and children were fetishes of mine at uni. With all the trials and tribulations here in recent months, I suppose Kendall won't attempt to begin those projects for a while. Although he probably will in time.*

'If my opinion means anything, you'll be a dedicated specialist in any field you undertake to study. Well, it's settled then. This week's been quite hectic. I'll be glad when it ends. Let's take a stroll.' Doctor Jarvis lowered his voice. 'I think it wiser to talk away from the house. Your mother listens to every word spoken on this front verandah. We cannot afford for her to suffer another relapse, not after the nasty shock of learning her husband was a war criminal. I prefer to be frank with you, Kendall. Soon you will have to face that fact, devastating as it sounds and will be. Personally speaking, I think Gigi has weathered this storm and come through without permanent mental damage. I expected her to have a coronary. My worse fear is she could be bordering on a stoke.'

Kendall cringed, even though he understood what Jarvis meant. 'Mum's not out of the woods yet, Stephen. I know how cantankerous she can be, if she's not well.'

'The results of her blood tests are back. They are negative. I've requested the lab technicians to recheck the test for Immune Deficiency, or AFV, as we term it. You understand *that* culture takes longer to process. Regarding your mother's smear test, I'll let you know the results, as soon as I receive their report.'

'It would be devastating to have to tell Mum that she's infected with gonorrhea or some other sexually transmitted diseases. She'd collapse, knowing that he was responsible for giving her a contagious illness.'

Jarvis concurred with Kendall. He knew his mother couldn't mentally cope after hearing the diagnosis of her contracting a horrible disease. 'Thus far none of my patients have contagious diseases, thank goodness.'

Stephen Jarvis hesitated to retrieve the car keys from his hip pocket. 'Kendall, I feel we haven't heard the last of Ian. He's a manipulative megalomaniac and an unscrupulous man. There's something in his past that will, in time be revealed. I'm certain of it. Keep in mind, it's your mother we must protect. She could still be vulnerable to any disease, if or when he does return home, given his sexual appetite for undesirable women. That's the main reason why she should take precautions. He will never change his sexual meandering whatever town or country he's in at present.'

'He will never sleep with mum again, Stephen. His disgusting habit of sleeping with prostitutes repels me. Even the thought of him touching my mother makes me cringe.'

'I feel the same as you. Your mother can't cope with any more worry Kendall. Still, I don't need to tell you that. Watch her closely and call me immediately if anything goes wrong, even if it seems trivial to her or you. It's the only way we'll be able to keep her health problems under control. Gigi is a gifted, strong-willed woman. Her frailty is being emotional. When stressed, she literally becomes a living time bomb.'

Kendall responded in a hushed tone as they neared the house. 'A week of rest without worry should make a huge difference to her health, in my opinion. Good night Stephen, I'll find it a challenge working with you, on my return from overseas.'

Jarvis waved from the car as he steered it around dust pockets in Abergeldie's dive. Kendall smiled with the prospect of a definite position in mind. *This news will please both my mother and Brianna. In the morning I'll mention his offer to Mum. She will criticise me because I haven't confirmed my holiday plans. Bridy knows I'll need something positive to look forward to, on my return from Switzerland and Austria in six weeks.*

Rain was an untenable topic to those attending the local show. The oppressive heat sparked fires which scorched the parched earth and inaccessible undergrowth. Dying crops brought to mind Dorothea Mackellar's poem *A Sunburnt Country*". From the pastoralists down to hardened bush-cockies, all sweltered while thinking of her famous words that no longer belonged to the past. Weather of this calibre proved to be a constant scourge to farmers and graziers struggling to keep food on the table in Nature's dust-laden country towns. The red-ochre soil lay open to the universe with little or no shade for sheep or cattle to huddle under from a burning summer sun. Clusters of trees in bare paddocks on both these properties provided little shelter for the men working in this sweltering, noon day heat.

Bjorn discussed the extreme weather pattern with his wife on their way home from the Cootamundra Show. Driving along what once was a scenic road, his thoughts lingered on their past success in the field of growing crops and his neighbour's failures.

'If other property owners haven't made provision in fruitful seasons for leaner times, they are fools in my opinion. In a way, I feel Gigi was right in giving up sewing canola and sorghum crops. The experiment proved a success while it lasted. Now she deserves a rest without a nagging husband ordering her to do his washing. Ian treated her worse than a common whore off the streets. Kirsten, I hope we never ever lay eyes on that bastard. He not worth worrying about. I'll kill him if he has the gall to show his ugly mooch here.'

Kirsten sided with her husband, to a point. She agreed with his idea of their neighbour being a murderer. 'Let's forget Ian Ross, or as we know him now as a Nazi criminal. War did dreadful things to innocent people, darls. I thank the Lord that you weren't killed flying those sorties over Germany and Poland. Gigi told me, she doesn't want Kendall to go anywhere near Auschwitz and those horrible places while he's overseas.'

'He won't, I can vouch for that. Now let me concentrate on driving this old bomb.'

Half a kilometre from home, his mind digressed to a time of plenty and Bjorn focused on his horses. *I consider myself fortunate because my first mare won the Canberra Stakes. In those days the glory of winning meant everything. Nowadays, I take it in my stride. If I win it's a bonus for spending time with my men, like Lanky, who teaches our young ringers the skill of handling and grooming foals and their sires or dams.*

'Darling, our sheep and cattle will never run short of lucerne, hay and other fodder under a hot summer sun.' Bjorn glanced in the side mirror to see if his horse floats and the cattle truck were following him. 'If our crops are harvested in their prime, in winter there will be enough to feed every animal on Jalna and Abergeldie. You know, I take pride in my work Kirsty.' About to alight and open their front gates, he looked across at his wife now dozing.

His response came as a yawn.

ACKNOWLEDGEMENTS

I wish to thank my sister Anne for her guidance to me through many trials. My niece, Suzie Wood for her letting me stay in her home in Bungendore when I needed help. Her mother, Julie Wood who read this book in its entirety in Batlow. A land holder and grazier who gave me an insight into sheep husbandry, Mr Richard Taubman lives in Young and he supplied me with the correct terminology on how to run and shear sheep. I found his help invaluable.

My colleagues at the Gold Coast Writer's group instilled in me the courage to continue writing when I was ill and found the work difficult to cope with on a constant basis.

I owe a huge debt to Mr Patrick Callahan who took a great interest in my talent and encouraged me to write novels related to my past history in varied fields of work I have been privileged to undertake in my life.

My doctors, past and present who inspired me to continue. My dear friends Dr. and Mrs. Deed and their family.

I also need to thank my publisher Andy McDermott of Publicious Book Publishing and his editor Mrs. Debbie Watson for their patience and encouragement in my times of trial.

* 9 7 8 0 6 4 8 1 2 5 5 2 5 *